Beyond competition: broadening the scope of telecommunications policy

Cammaerts B. & Burgelman J.C. (eds)

Beyond competition: broadening the scope of telecommunications policy

VUB UNIVERSITY PRESS

Media en Maatschappij - 3

Cover: Danny Somers
Layout: Boudewijn Bardyn

Waversesteenweg 1077, 1160 Brussels - Belgium
Fax: ++ 32 2 629 26 94
e-mail: vubpress@vub.ac.be
ISBN 90 5487 267 5   NUGI 656
D / 2000 / 1885 / 015

Table of Contents

Acknowledgements

Research on the implications of the information society for governments, regulation and policy presented in this book is supported by DWTC, the Belgian Federal Institute for Science Policy. In that respect we would like to thank all our academic partners for their contributions and help. Our SMIT colleagues deserve, as always, credit for their logistical help and intellectual support. We would also like to thank all the participants of the seminar 'paving the way for a new public telecommunications policy' for their active contribution to the discussions held that day. We are especially grateful to the different publishers who granted permission to reprint; Sage, Rowman & Littlefield, SERDI and the Economic Research Foundation for Communication. Finally we thank Geert Frateur for his revisions and the VUBPRESS for their patience.

Bart Cammaerts
Jean-Claude Burgelman

© Chapter 6 was previously published in Picard, R.G. (eds): (1998) *Evolving Media Markets: Effects of Economic and Policy Changes*, pp. 57-80 and is reprinted with the acknowledgment of the Economic Research Foundation for Mass Communication, Turku - Finland.

© Chapter 9 is due to be published in *Media, Culture & Society* (1999), Vol.21 and is reprinted with the acknowledgment of Sage, London - UK.

About the authors

Pieter Ballon studied modern history at the Catholic University of Leuven and information science at the University of Antwerp. He previously worked as a contract researcher for SMIT (Free University of Brussels) and ASCoR (University of Amsterdam). He is now a junior researcher and consultant at the Centre for Strategy, Technology and Policy (STB) of the Dutch research institute TNO. (ballon@stb.tno.nl)

Paul Belleflamme holds a Ph.D. in Economics from the FUNDP (Facultés Universitaires Notre-Dame de la Paix), in Namur (Belgium) and a Master in Economics from Columbia University, New York. Until August 1999, he worked as a researcher for CITA (FUNDP-Namur). He is now lecturer in Economics at the Queen Mary and Westfield College, University of London. (pbe@info.fundp.ac.be)

Marc Bogdanowicz holds a Bachelor's degree in Education Sciences and a postgraduate degree in Organisational Sociology from the University of Liège (Belgium). He directed for 8 years an academic consultancy and adult training centre before joining the Evaluation and Technology Assessment Department of LENTIC. (M.Bogdanowicz@ulg.ac.be)

Jean-Claude Burgelman is Professor in National and International Communications at the Free University of Brussels and guest professor at UNISA - South Africa in the post-graduate programme in Telecommunication and Information Policy. He has a Ph.D. in Communication and a Master's in Science and Technology Dynamics, both from the Free University of Brussels. He directs the centre for the 'Study of Media, Information and Telecommunication' (SMIT). He is currently on leave at the Institute for Prospective Technological Studies, JRC (Sevilla) as a visiting scientist. (jcburgel@vub.ac.be)

Bart Cammaerts graduated as a social worker in 1992 and as a political scientist in 1997 (Free University of Brussels). He worked as a journalist in Latin

America and as a press-secretary and advisor on Information Society policy to the Belgian vice-prime minister, minister of economic affairs and telecommunication. Currently he is researcher at SMIT and in charge of the IUAP-project. (Bart.Cammaerts@vub.ac.be)

Jean-Marie Cheffert is Professor at the Faculty of Law and at the Faculty of Economics of the FUNDP (Facultés Universitaires Notre-Dame de la Paix-University of Namur, Belgium). He holds a Ph.D. in Economics from that university and two Masters in Economics (University of London (LSE), University of Namur). Since 1992, he has been Director of the postgraduate diploma in Information Technologies Law (DGTIC) in Namur and head of research at CRID. (jean-marie.cheffert@fundp.ac.be)

Nicholas Garnham studied English literature at Cambridge University and then worked in the U.K. television industry as a documentary director. He has worked at the University of Westminister since 1972, where he is Professor of Media Studies. He is a founding editor of the journal Media, Culture, and Society. (garnhan@westminster.ac.uk)

Anders Henten has an MA and Ph.D. in Communications and International Development studies. He is currently associate professor at the Centre for Tele-Information (CTI) at the Technical University of Denmark. (henten@cti.dtu.dk)

Claire Lobet-Maris is a sociologist and holds a Ph.D. in Managerial Sciences (Université Catholique de Louvain - Belgium). She is Professor at the Computer Sciences Department of FUNDP and co-director of the 'Cellule Interfacultaire de Technology Assessment'. Until recently she was in charge of research and development policy in telematics and multimedia for the regional government of Wallonia (clo@info.fundp.ac.be)

Robin Mansell is Professor of Information and Communication Technology Policy at SPRU - Science and Technology Policy Research, University of Sussex, UK. She co-directs the SPRU Information, Networks & Knowledge (INK) research centre. (R.E.Mansell@sussex.ac.uk)

Denis McQuail is Professor emeritus of the University of Amsterdam and curently associated with the Amsterdam School of Communication Research. He is also Visiting Professor at the University of Southampton. England. He has published widely on aspects of media theory, research and policy. (106233.3000@compuserve.com)

David Neice is a doctoral research student at SPRU - Science and Technology Policy Research, University of Sussex. He was a senior research analyst for many years with the Canadian government on information society policy (D.C.Neice@sussex.ac.uk)

Yves Poullet is Professor at the Law Faculties of the University of Namur, director of CRID (Centre de Recherches Informatique et Droit), member of the Belgian Privacy Commission and chairman of the Legal Observatory Board Task Force on Access to public electronic documents and electronic democracy. (yves.poullet@fundp.ac.be)

Jan Servaes is Dean of the Faculty of Social and Political Sciences, Professor and Chair of the Department of Communication at the Katholieke Universiteit Brussel. He is Director of the Research and Documentation Centre 'Communication for Social Change' (CSC), and President of the Participatory Communication Research Section of the International Association of Media and Communication Research (IAMCR). He has been teaching at the Universities of Cornell (Ithaca, USA), Nijmegen (The Netherlands), Thammasat (Bangkok, Thailand), and Antwerp (Belgium). (Jan.Servaes@kubrussel.ac.be)

W. Edward Steinmueller is Professor of the Economics of Technical Change and Innovation at SPRU - Science and Technology Policy Research, University of Sussex where he co-directs the Information, Networks & Knowledge research centre and is responsible for doctoral student admissions. (w.e.steinmueller@sussex.ac.uk)

Marie d'Udekem-Gevers has a Ph.D. in anthropology (Université de Liège), a Ph.D. in Zoology (Université Catholique de Louvain) and a Master in Computer Sciences (FUNDP). She is currently a researcher at the Cellule Interfacultaire de Technology Assessment (CITA) of the FUNDP (mgevers@info.fundp.ac.be)

Leo Van Audenhove has a master degree in communication from the Free University of Brussels. Currently he is doing research on the African Information Society at SMIT (lvaudenh@vub.ac.be)

Béatrice van Bastelaer has a Master's Degree in Economics from the University of Namur (Belgium). Since 1992, she has been working at CITA (*Cellule Interfacultaire de Technology Assessment*) at the University of Namur. CITA is a research unit specialised in the assessment of information and communication technologies. (Beatrice.vanBastelaer@info.fundp.ac.be)

Jan Van Cuilenburg is Professor of Media and Telecommunications Policy and Management at the University of Amsterdam. He is Dean of the The Amsterdam School of Communications Research ASCoR. He is also Director of the Netherlands Press Foundation, member of the Netherlands Media Council and the Netherlands Advisory Committee on Post and Telecommunications. (vancuilenburg@pscw.uva.nl)

Eric Van Heesvelde has a law degree from the University of Ghent, where he is currently guest professor. He once was a legal advisor for a private company before joining the Belgium Planning Office where he was responsible for the sector of communications, transport and public investments. He also served as head of the cabinet of the Minister of Science Policy and vice-head of the cabinet of the Minister of Cooperation & Development. Now he is General Administrator of the Belgian Institute for Post and Telecommunications (BIPT). (eric.van.heesvelde@bipt.be)

Pascal Verhoest holds a Ph.D in social sciences and now is senior researcher and advisor of the Netherlands Organisation for Applied Scientific Research (TNO) – Institute for Strategy, Technology and Policy (STB), where he is responsible for the telecommunications Research Unit. He also teaches at the Katholic University of Brussels (KUB) and the Free university of Brussels (VUB). (verhoest@stb.tno.nl)

Jean-Claude Burgelman & Bart Cammaerts

Introduction

This book is the result of the seminar *Paving the way for a new public telecommunications policy*, which SMIT-VUB organised at the end of 1998 within the framework of the IUAP-project (Inter University Attraction Poles) *Research and Interdisciplinary Evaluation on the Information Society: Networks, Uses and Roles of the State.*

IUAP's are networks of centres of excellence, funded by the Belgian Federal Institute for Science Policy, DWTC. This IUAP started in 1997 and, besides SMIT (VUB Brussels), also comprises CRID/CITA-FUNDP, which is the project leader (Namen) and Lentic from Ulg (Luik).

The aim of this particular IUAP network is to set up a pole of research and interdisciplinary assessment on the development of the information society, and in particular of the information superhighways in Belgium. It also aims to increase basic knowledge about the impact of information technology on society. Through both conceptual studies and empirical analysis, IUAP wants to contribute to the formulation of sustainable development policies for the Belgian information society.

Research is structured around three main axes, each one including a conceptual, an empirical and an evaluative/prospective dimension.

Axis 1. Economy of innovation and development of networks
The first work axis deals with the identification of development scenarios for communication networks of the future. Its conceptual dimension aims at setting up an analytical framework enabling the study of the different existing technical solutions for communication networks. This task starts from an analysis of the interactions between and the strategies of the actors in the markets for telecommunications and information and communication technologies (ICT). These observations will be confronted with economic theories related to innovation, investment, standardization and industrial organisation. The empirical dimension consists in drawing up a map of the actors,

undertaking detailed analysis of their interests, of the means used by them, and of the strategies of alliance and conflict in which they engage. Finally, the evaluative/prospective dimension aims at establishing and assessing different existing scenarios, with special attention being brought to the interests and controversies at stake.

Axis 2. Dynamics of uses and logics of appropriation
The objective of the second axis is twofold: first, the need to understand more profoundly how ICT are used in their social, cultural and organizational contexts. Such better understanding must lead to the identification of the conditions under which the demand for innovative services may arise and develop in the information society. In the conceptual dimension, the associated teams take advantage of their previous work and set up a common framework to understand the dynamics behind the usage of ICT, through the analysis of social appropriation behaviour, demand formation, and social as well as cultural transformations. The empirical dimension deals with the observation of some communities of users (or creators) of networks (or of communication services) which serve as prototypes for the communication systems of the future (Internet, inter-organizational communication systems, etc.). Finally, the evaluative/prospective dimension implies a definition of relevant action schemes in the field of formation (about new cognitive models, for instance), in the conduct of social experiments and in regulation. This dimension also consists in defining the role of the State and its expected actions concerning the promotion and regulation of innovating uses.

Axis 3. Management of the public/general interest and State intervention in the information society
The transformation of society into an information - and thus knowledge based - society prompts a multitude of questions hitherto unanswered. The third axis has the ambition to answer at least some of those questions through fundamental scientific research. This axis focuses on the role of the state in the development towards an information society and its role as protector of the 'public interest'. The conceptual dimension aims at clarifying the current debate on the 'public service' notion through work on the socio-historical evolution of the public/general interest notion, on the role of the State, and on the ethical foundations capable of taking into account cultural differences and assisting regulatory authorities. The empirical dimension implies an analysis and comparison of various existing approaches, centred around defining public/general interest, its subsidisation, and the regulatory role granted to the States. Finally, the evaluative/prospective dimension seeks to

determine the economic feasibility and social relevance of the different approaches to the public/general interest standard and propose a novel framework for State intervention in the information society. This raises the issue of the relationship between self-regulation and the ethical considerations regarding public/general interest regulation in the information society.

SMIT directs research in this last axis and the above mentioned seminar was organised to explore in depth some critical issues that ongoing research within IUAP had already brought to the forefront. The seminar also served to confront IUAP research efforts with international scholars.

Hence 62 participants, coming from industry, academia and the private sector attended the seminar (coming from 6 countries). Three international renowned speakers introduced the seminar (Prof. N. Garnham, Prof. R. Mansell and Prof. B. Melody). After this plenary session three workshops were organised: (1) Political Perspectives; (2) Social Regulation; (3) Market Regulation. Members of the IUAP-network presented the bulk of the papers in these workshops.

The seminar was constructed around the central question how public policy has to be adapted in order to reconcile market and social objectives in a global, privatised and competitive communication market.

As Collins (1996) points out, this balancing exercise is as such nothing new. It has always been the goal of public policy in Western democracies with regard to broadcasting and telecommunications to achieve a balance between the interests of the private sector and the more general interests of social well-being. In Belgium (Burgelman, Punie & Verhoest, 1995b) as like in most European countries, this resulted in a strong public service broadcasting institution and a de facto or de jure monopoly in public telecommunications operators. This searching for a balance has in fact been the main imperative for building what has been labelled the welfare state.

However, if history always repeats itself, it never does in the same way. This goes for policy in the realm of communications as well. Indeed, the balances to be looked for today, a major theme in the articles of part 1 (political perspectives) of this book, are quite different than those that had to be struck in the past. This is mainly due to two sets of reasons.

On the one hand, as is explored in part three (market regulation), because the parameters on which the balances are built appear to be quite different from those on which the past balances were achieved. There are no longer technical reasons like network integrity that necessitate imposing some kind of monopoly, traded off by some kind of universal service provision. Good regulation can indeed realise universal service without going against competition which in itself and until now has proven to boost the efficiency of the sector. At least for the business user. However, to which extent these benefits will also last - in view of the huge concentration movements going on - and will tickle down to the rest of the population is yet unclear and at the locus of policy concerns.

A debate that isn't so easy to conclude if one takes into account that, on the other hand, telecommunications policy is, and certainly will become, much more than the roll out of a copper cable. The articles of part three all clearly indicate that in the future telecommunications policy will be a key element of an integrated policy towards an information society. What this information society will be - networked society or knowledge society – is still unclear. But it is crystal clear that access to technologies which in turn gives access to knowledge and information, as well as the skills to master them, will also become core issues for a new public policy.

Given the growing importance of communication and networks for all domains of society - from broadcasting over business and health care to education and internet games - it is clear that in the next century public policy with regard to telecommunications will be much more than what it was up till now.

That doesn't mean that every activity which will run over a network and hence becomes tele -whatsoever, has to be seen as an integral part of future telecommunications policy. In fact, this 'holistic' approach in which the whole society is going on-line and hence subject to communications policy, is, though popular in some textbooks and certainly in the hype literature - to be or not to be digital - rather ridiculous. It does mean, however, that new balances between telecommunications policy and other policy domains like education, welfare, medical health, business and so on, have to be traded off in a synergetic way.

Therefore, as underlined in the work of Mansell and When (1998), good policy with regard to the building of an information society must rely not only on sufficient technical and material resources (the networks) and skills, but also be coherent with other societal policies. The latter point in particular opens up a wide range of questions. Because not only did the internal parameters of telecommunications policy change, as outlined above, but the whole context against which this policy has to be defined changed as well.

Let us give an example: the social network of insurances and securities that encaptured most of Western Europe's poor is in the process of slowly being dismantled. New jobs are indeed created and thus soaking part of the unemployed, but unskilled labour is declining as well and the new jobs require new skills which these poor certainly do not have and will probably never acquire on their own. Hence, the information society also leads to the persistence and even re-enforcement of structural unemployment by network-based occupations. As observed by Freeman and Soete:

> 'The advent of the information society has thus been accompanied by a reversal of all those trends towards social justice and improved welfare

> services, which were such a characteristic feature of the quarter century following the Second World War. It seems probable that some features of the information society will exacerbate these trends even further. In particular a division is taking place between the "information rich" and the "information poor"'. (Freeman & Soete, 1997: p. 410).

How to contribute to a solution for this problem, will probably be one of the biggest policy challenges in telecommunications for the next century.

We hope that this book helps move the debate further on into such a broader approach. Because it was indeed felt by the organisers of the workshop that too much debate on telecommunications policy sticks to issues such as interconnection, fair competition and so on. Important as such, but quite trivial in the long run. A perspective that underlies precisely the work SMIT is carrying out in the third pole of the IUAP.

As stated before, this book tries to mirror as much as possible the seminar *Paving the way for a new public telecommunications policy*, which SMIT-VUB organised at the end of 1998. As the attentive reader will notice, the contents of the book diverges a little from what actually happened during the seminar. Some authors submitted another article than the paper presented, which explains why the synthesis of the debates held within the different workshops might be somewhat different. Two articles were added (see the conclusion) in order to put the debate on telecommunications policy in the perspectives of the Belgian market in particular and innovation policy in general.

PART I

Political Perspectives

Bart Cammaerts, Leo Van Audenhove

Introduction

A new Social Contract for the Information Society

The liberalisation and privatisation of public services prompts new and urgent questions about the public interest values inherent to services such as energy, transport and communication. In this book the discussion is limited to information and communication technology services. ICTs have become so central in our post-industrial societies that they touch every aspect of what became the welfare state. A welfare state which in itself is in crisis due to a complex of social, political and economic transformations. The latter referring to increased mobility, changing patterns of industrial strategy and the flexibility of labour, the former to the liberalisation of public services, demographic shifts and changing patterns of governance.

These evolutions lead to a renewed interest in the concept of the welfare state. Scholars from different scientific backgrounds have over the last years contributed to the field and even elaborated new inter-disciplinary approaches. We now dispose of a sound and extensive framework to analyse welfare state regimes in an historical and comparative perspective (Esping-Andersen, 1990; Castells & Mitchell, 1992; Jones, 1993; Ferrera, 1996).

During the eighties, conservative and neo-liberal thinkers questioned the welfare state (Barry, 1997; Wilding, 1997; Denham, 1996). As we are nearing the end of this century minds are changing in many respects. The laws of Foulcault's pendulum have apparently not been affected by recent evolutions. Many factors indicate that the welfare state concept is far from dead, on the contrary. International institutions such as the Worldbank or the OECD have, after years of neglecting social aspects of development, put the notion of welfare back on their policy agendas (Nulens & Van Audenhove, 1998; 10, OECD, 1998; chap. 2). In the political field, in thirteen out of the fifteen countries of the EU, social democrats — the historical architects and defendants of the welfare state — have taken over political power after years of conservative or neoliberal rule. In spite of globalisation, thatcherism and reagonomics, many of the European welfare state regimes remain rather strong (Pierson, 1994). However, it remains unavoidable and essential that the nature of the welfare state is reviewed and adapted to the (new) needs of the post-industrial society and its citizens (Group of Lisbon, 1994; Culpit, 1992; Esping-Andersen, 1996). As the nature of the nation state is changing due to growing political and economic interdependence, budget constraints, privatisation, and anti-state rhetoric, it is essential and unavoidable that we review and adapt the nature of the welfare state. Conservatism on both sides of the political spectrum has hitherto prevented the formulation of innovative answers to the crisis of the welfare state.

As new media and ICT services increasingly play an important part in the development of new needs with respect to the welfare of citizens, it is only logical that they should be part of a broad revision of the welfare state. The role of ICTs in job destruction (creation) and education, in the modernisation of administrations, in the medical services, et cetera make that access to these services and the capability to use them became of prime importance for every citizen wanting to participate economically, socially, politically and/or culturally in our post-industrial or so-called 'information society'. The current restructuring of the (tele)communication sector by way of the introduction of competition should therefore be complemented with a thorough rethinking of public policy towards services such as ICTs. It is for instance unacceptable that access would only depend upon a person's financial resources. In this regard (social and market) regulation remains essential. Even if we assume that competition as such could have a regulatory effect, we must accept that a perfect competitive market never existed and probably never will. Market distortions such as oligopolies, cartels, monopolistic competition, regulatory capture, — so characteristic for the telecommunications and ICT markets — undermine public interest objectives (Melody, 1997b; 498). It becomes apparent that besides access, protection of consumers and citizens will become another major challenge for the new welfare state.

Considering all this, it seems imperative to widen the scope of (tele)communication policy in terms of other policy areas; e.g. social policy, educational policy, economic policy. In this respect the classic dichotomy between state and market must be overcome. Simplistic analyses such as 'the state is concerned with the public interest and the private sector with profit maximisation' or 'the state is inefficient and should leave everything to a well-performing and flexible private sector' are no longer valid. Telecommunication history shows that state-owned monopolies are not as such a condition for a universal, qualitative and cheap service (Verhoest, et.al., 1991). On the other hand neither are liberalisation and privatisation. Competition on the European residential market is rare, prices stay high, often with the consent of the regulator in order to compensate lost international traffic and big users (Galal, et.al., 1994; 99, Joosten, 1991; 35). However, the divide between social and economic goals often is not as deep as perceived. Why should more welfare not be in the interest of the market? And if so, why would the market not contribute to more welfare? This logic can be turned around. Why should a prosperous economy not be in the public interest? Without wealth creation it is not possible to redistribute it. Essential here is the premise that the economy stands at the service of people and not the other way round. If we accept this, we also see that welfare is in the interest of both state and market.

Therefore, it seems that with the turn of the century at hand the time has come to reformulate the so-called social welfare contract and to include new rights such as access and protection to citizens and consumers alike. Historical evidence shows that if this reformulation is to be successful, it will have to be done on the basis of a consensus between the public and private sectors (Burgelman, 1997a: 137). The social contract for the industrial society was also a compromise between labour and capital. Moreover, due to political and economic transformations, the state is no longer able to guarantee every right on its own, it needs all other actors. The lack of a social welfare contract for the post-industrial society in combination with growing interdependence leads towards a legitimacy crisis and the strengthening of dualisms within our society which in turn might affect the potential for long-term economic growth. The knife cuts both ways. For the time being, human sciences fail to take a stance in this often normative debate. More research should be done into which social policies work and are worth pursuing and how they can be achieved. The aim of this book is to start the quest for some of the answers to these questions.

Nicholas Garnham

Amartya Sen's 'Capabilities' Approach To The Evaluation Of Welfare: Its Application To Communications

To concern ourselves with the Welfare State is to concern ourselves with entitlements. The theory and practice of the welfare state is based upon the assumption that all citizens have entitlements to certain forms of welfare; that the interaction of private individuals pursuing their own interests, particularly their own economic interests, within civil society will not ensure that all citizens receive these entitlements; and therefore the state has a responsibility for the delivery of such entitlements. Whether these entitlements should be equally distributed among citizens or whether a basic level should be guaranteed is a matter of subsidiary debate.

But the fundamental question remains entitlement to what? This is not only of obvious general importance. But whether and if so in what form we can include communications within the remit of the welfare state will depend upon our answer.

It has been the major strength of Amartya Sen's work to focus on this issue. Sen starts from an egalitarian perspective. Thus he phrased the question 'Equality of What?, the title of his celebrated 1979 Tanner Lecture. But before I turn to Sen's argument I think we need to distinguish between types of State intervention and the reasons for them.

Much State intervention in the economy has nothing to do with welfare as I have defined it but is a response to market failure in a narrower sense. Thus in the field of communication we have to distinguish between policies based on the assumption that it is more efficient to produce a good or service via a regulated monopoly from those whose aim and justification is the general provision of what is considered to be an essential service. This issue is raised both in telecommunications under the rubric of universal service and in broadcasting under the rubric of public service. Confusion in this area is exacerbated by the language we use, since market failure arguments may be couched in the language of welfare economics as welfare losses. Thus for instance a flat-rate license fee system for financing broadcasting can be justified either in terms of welfare maximisation because of the zero marginal cost of reaching extra viewers or in terms of the well-being to be derived from the wider range and quality of programming and the democratic possibilities to be derived for a common service unbeholden to commercial forces. Similarly, rate averaging and regulatory prohibitions on cream skimming to ensure maximal network penetration, are different from arguments based upon the need of citizens to have a telephone for full participation in social and economic life and therefore, from a welfare perspective, their right to such a phone, however in fact provided.

Sen's approach to the question of defining and measuring welfare, well-being or what his group now calls the quality of life, is immeasurably sharpened by starting from a normative egalitarianism. As he puts it:

> 'Every normative theory of social arrangements that has at all stood the test of time seems to demand equality of something. [...] Equal consideration at some level a level that is seen as important is a demand that cannot be easily escaped in presenting a political or ethical theory of social arrangements. It is also of considerable pragmatic interest to note that impartiality and equal concern, in some form or other, provide a shared background to all major ethical and political proposals in this field that continue to receive argued support and reasoned defence' (Sen, 1995: 12-19).

He goes on to argue that arguments against any form of equality generally take the form of arguing for equality of something else, e.g. liberty as opposed to income. This leads him to stress the importance of being clear about the Space within which we are arguing for equality. Here we can see some similarity to Walzer's notion of Spheres of justice. Sen's argument leads him, as we shall see, to argue that policies on distribution and its measurement should focus not on income, or what both Rawls and Dworkin call resources, but on what he calls 'The Space of Functionings,' defined as the various things a person may value doing (or being). He goes on to argue that one can then focus on either a) realised functionings (what a person is actually able to do) or b) on the set of alternatives she has (her real opportunities). These alternatives or potential choices he dubs 'Capabilities'. This concept of capabilities derives from the Aristotelian concept of 'dunamin' meaning 'capability of existing or acting.' In the last analysis Sen is advocating the Space of Capabilities as the appropriate space within which egalitarian welfare policies should be designed and evaluated. This may at this point appear very abstract but it leads to radical, positively discriminatory conclusions. People differ in their natural endowments, for instance, some may suffer from physical disability. Those so disadvantaged may require a significantly higher level of resources or income to produce the same level of capability. In the field of communications, it leads to the conclusion that it is not access in a crude sense that is crucial but the distribution of the social resources which make access usable. At a simple level additional support for the blind or deaf, but at a more fundamental level attention to levels of educational inequality which determine the different utilities that can be gained from a given level of supply.

Before returning to these implications of Sen's position within the field of communication, let me go through the steps in the argument by which Sen arrived at these conclusions.

Much of Sen's argument may appear very theoretical and abstract, but it is important to stress that he is in the end concerned with pragmatic questions of policy assessment and with the measurement necessary for such assessment — for instance the measurement of poverty or of the relative development levels of different countries. But he stresses that to make appropriate assessments and develop appropriate metrics, requires us to decide the value space within which our assessments and measurements will be made.

The capabilities approach derives from a critique of, on the one hand, welfare economics and the utilitarianism that underlies it and on the other of Rawls' *Theory of justice* and its concept of primary goods.

Sen criticised the welfare approach on the grounds that, both for hedonistic and preference satisfaction, welfarists focus 'not on the person's capabilities but on his mental reactions.' This, he argued, was an unsuitable guide to policy if only because a person may adjust her expectations to her condition.

As he puts it in *Inequality Re-examined:*

> 'The utilitarian notion of value, which is invoked explicitly or by implication in much welfare economics, sees value, ultimately, only in individual utility, which is defined in terms of some mental condition, such as pleasure, happiness, desires [...]. There are different problems with different interpretations of utility, but they share the programme of getting the evaluation done indirectly through using some psychological metric like happiness or desire. This is precisely where the main difficulty lies. While being happy may count as an important functioning, it cannot really be taken to be all there is to leading a life (i.e. it can scarcely be the only valuable functioning). [...] The problem is particularly acute in the context of entrenched inequalities and deprivations. A thoroughly deprived person, leading a very reduced life, might not appear to be badly off in terms of the mental metric of desire and its fulfilment, if the hardship is accepted with non-grumbling resignation. In situations of long-standing deprivation, the victims do not go on grieving and lamenting all the time, and very often make great efforts to take pleasure in small mercies and to cut down personal desires to modest — "realistic" — proportions. Indeed, in situations of adversity which victims can-

> not individually change, prudential reasoning would suggest that the victims should concentrate their desires on those limited things that can possibly be achieved, rather than fruitlessly pining for what is unattainable. The extent of a person's deprivation, then, may not at all show up in the metric of desire-fulfilment, even though he or she may be quite unable to be adequately nourished, decently clothed, minimally educated, and properly sheltered. The problem of entrenched deprivation is particularly serious in many cases of inequality. It applies particularly to the differentiation of class, community caste and gender. While the nature of these deprivations can be brought out more clearly by concentrating on socially generated differences in important capabilities, some of that gain would be wasted if the capabilities themselves were to be assessed, after all, in the metric of utilities' (Sen, 1995: 54-55).

I would note in passing the overlap between this approach to deprivation and that found in Bourdieu's sociology and particularly in the mobilisation of the concept of habitus to analyse the ways in which socially determined expectations are literally incorporated — that is to say inscribed in our very bodies. An important strand in Sen's subsequent work has been the analysis of the ways in which deeply entrenched forms of gender discrimination have produced, in situations of measurable absolute and relative deprivation, just this sort of acceptance among many women.

Given this critique it is hardly surprising that Sen was drawn to Rawls' theory of justice as an alternative way of thinking about inequality for Rawls' theory of justice is built upon a critique of utilitarian theories of distributive justice and a move away from mental states as the appropriate space of evaluation. In particular, Rawls argued that any theory of justice has to confront the problem of opulence. How do we weigh, using a purely mental or hedonistic metric, the pleasure gained by the poor person from having just enough to eat with that gained by the rich person from consuming caviar or fine wine. Rawls' answer was to move from mental states to actual goods, what he called primary goods — the basic resources necessary for the free pursuit of happiness — which he argued should be equally distributed. Against this Rawlsian position Sen argued that differently constructed and situated people require different amounts of primary goods to satisfy the same needs. 'What people get out of goods depends on a variety of factors, and judging personal advantage just by the size of personal ownership of goods and services can be very misleading. It seems reasonable to move away from a focus on goods as such to what goods do to human beings' (Sen in Nussbaum & Sen, 1993). He gives as an example, given an equal bundle of goods, the comparative advantage of a soundly limbed person over a paraplegic in the exercise of those freedoms of which it is the aim of Rawls' theory of justice to ensure equality.

Sen then went on to argue that what we should be concerned with was not mental states or goods but what he called 'functionings', which he defined as the various things a person may value doing or being.

The functionings relevant for well-being vary from such elementary ones as escaping morbidity and mortality, being adequately nourished, having mobility etc. to complex ones such as being happy, achieving self-respect, taking part in the life of the community, appearing in public without shame. The claim is that the functionings make up a person's being, and the evaluation of a person's well-being has to take the form of an assessment of those constituent elements (Nussbaum & Sen, 1993: 37).

It is important for the next step of the argument that Sen makes a distinction between states of well-being on the one hand and on the other agency, or the achievement of states of well-being. This is important for two reasons. First, because the enjoyment of valued well-beings may either not have been achieved or indeed be individual by individual action. He wishes to avoid the Rawlsian stress on the individual pursuit of well-being and the danger of lapsing into a merit as opposed to need-based approach to the assessment of equality or entitlement. As he has put it:

> 'A system based on needs would seem to have greater use for the complex idea that we call humanity. Even for the limited application of the merit principle it can be argued that the measure of merit is culture specific. [...] Merit is a bit of an accident not only in its origin, but also in its being treated as merit' (Sen, 1997: 105).

For instance, freedom from malaria may be an extremely highly valued and basic functioning but it cannot usually be said to be the result of the agency of the individual who enjoys it. The second reason however is that he also regards agency as itself a valued functioning. And it is for this reason that he moves from functionings to capabilities as the basic evaluative space. Functionings are what a person does or is. Capabilities are the set of alternative functionings she has (her real opportunities). He does not deny that the measurement of achieved functionings maybe important in assessing well-being and indeed be easier to compare than capabilities. But that the distinction is important in both ethics and public policy he illustrates, in a field he has made very much his own i.e. the study of famine, by the distinction we make in ordinary language between starvation and fasting. Starvation is an absence of the functioning of nutrition which is not chosen by the agent. He or she does not possess the capability of nutrition. Fasting on the other hand may be a valued state, but it has to be chosen by someone who also has the option to eat. It can thus only be the action of someone whose capability set includes eating as well as not eating as possible functionings. Death and suicide would be another example. This move from achieved functionings to capabilities as the evaluative space for entitlements is particularly important in the context of the critique of welfare as a form of forced consumption, since in the capabilities approach such cases of forced consumption, whether the force derives from market relations or the state, can be seen, in comparison with the achievable relevant capabilities, as a form of deprivation.

Having established that welfare is concerned with the distribution of and entitlement to capabilities and functionings rather than either resources or utilities, we are still left, for the purposes of ethical and policy evaluation, with the problem of hierarchising these functionings and capabilities. As Sen puts it:

> 'There are always elements of real choice regarding the functionings to be included in the list of relevant functionings and important capabilities [...] There is no escape from the problem of evaluation in selecting a class of functionings and in the corresponding description of capabilities. [...] It is certainly clear that some types of capabilities broadly conceived are of little interest or importance, and even ones that count have to be weighted vis-à-vis each other. But these discriminations constitute an integral part of the capability approach' (Sen, 1995: 44-45).

How does Sen approach this problem? He firstly argues that since functionings are robustly heterogeneous the need to weigh them against one another cannot be avoided. But he also argues this is true of other approaches since heterogeneity of factors that influence individual advantage is a pervasive feature of actual evaluation. For instance, an approach based on relative income:

> 'simply assumes that there is something homogenous called "the income" in terms of which everyone's overall advantage can be judged and interpersonally compared [...] this does not resolve the problem only evades it' (Sen, 1997: 204).

How then are the weights to be selected?

> 'This is a judgmental exercise and it can be resolved only through reasoned evaluation. In making personal judgements, the selection of the weights will be done by a person in the way she thinks is reasonable. But in arriving at an "agreed" range for social evaluation (e.g. in social studies of poverty), there has to be some kind of a reasoned "consensus" on weights (even if it is of an informal kind). While the possibility of arriving at a unique set of weights is rather unlikely, that uniqueness is not really necessary to make agreed judgements in many situations, and may not indeed be required even for arriving at a fully complete ordering.'

For Sen, and here he shares a position similar to Rawls, a very important part of the process of arriving at these comparative weights is a Kantian move to place oneself in the place of the other. When making comparative evaluations of the functionings and capabilities of others the evaluator needs to place himself in the position of the other and ask whether with that capability set they would regard themselves as absolutely deprived and/or relatively disadvantaged. Thus functionings and capabilities are comparatively evaluated and 'in a democratic context, values are given a foundation though their relation to informed judgements by the people involved.' Thus people living in a given society rank, through a process of reasoned argument, the functionings and capabilities of morbidity vis-à-vis mortality, nutrition vis-à-vis mobility, morbidity vis-à-vis participation in social life, mobility vis-à-vis communication, such that they can at least produce partial rankings which, for instance, agree that nutrition is more important (more basic) than communication and morbidity than participation in social life. In the end, Sen argues: 'non-arbitrariness of valuation in a democratic society' depends upon 'openness to critical scrutiny, combined with — explicit or tacit — public consent.' In particular 'the metric of market exchange cannot be used to weigh different functionings any more than different commodities — because it tells us nothing about interpersonal comparisons' (Sen, 1997: 206). The use of such a metric is thus precisely a way of avoiding the need for open critical scrutiny of the weights being used.

This approach to relative weighing has been criticised from a classic liberal perspective on the grounds that, once one accepts, as Sen does, the importance of agency or freedom as an aspect of well-being, the diversity of human goals makes relative weighings in practice impossible and from an Aristotelian perspective, for instance by Martha Nussbaum, that one's dilemma can only be avoided by developing 'an objective normative account of human functioning.'

Sen is clearly torn. On the one hand, his move away from the space of utility and resources to that of functionings and capabilities is a move, as he himself admits when pointing to the origin of the concept of capability in Aristotle's notion of dunamin, towards a space of evaluation within which objective assessments of relative advantage, suitable as a foundation for egalitarian public policy can be arrived at by reasoned argument. On the other hand, he sides with the liberals in seeing freedom itself as a valuable functioning, indeed that which makes capabilities rather than mere functionings the ultimate space of evaluation. His response is that on the one hand the capabilities approach is superior to a commodities or income-based approach for the interpersonal and inter-social comparisons essential to the justification of entitlements since in actual practice, while commodity requirements may differ as between individuals and societies the underlying capabilities are much less varied and are close to what Aristotle called 'non-relative virtues'. On the other hand, he argues that the first priority is to identify the appropriate space of evaluation

— that one can agree to focus on capabilities while disagreeing both on the exact grounds underlying the determination of relative weights and on the actual relative weights chosen. Even within the agreed value space of capabilities there can be different value purposes, in particular the relative weight given to well-being and agency. (Note this is not the same as the distinction between negative and positive freedom).

Having hopefully cleared the ground, let me now turn back to the question of communication entitlements within a capabilities approach. Sen is fond, if only, I suspect, to annoy the neo-classics and neo-liberal pro-marketeers, of quoting Adam Smith on necessaries:

> 'By necessaries I understand not only the commodities which are indispensably necessary for the support of life, but whatever the custom of the country renders it indecent for creditable people, even the lowest order, to be without' (quoted in Sen, 1981: 18).

And coupling it with Marx's argument that while 'a historical and moral element' enters the concept of subsistence, 'nevertheless, in a given country, at a given period, the average quantity of subsistence necessary for the labour is practically known' (quoted in Sen, 1981: 18). Sen thus argues that it is possible to arrive at an objective description of such necessary subsistence in a given society at a given time and that such description must be firmly distinguished from any prescriptive statement. A major task for communication scholars, it seems to me, is to provide just such a description in the realm of communications.

Let me now turn to the implications of Sen's capability approach in the field of communication. I imagine that we can all agree that in all modern societies to varying degrees the functioning of social communication is a part of well-being. Certainly it can be reasonably argued that, in developed countries at least, it is indecent, in Smith's sense, for any citizen to be without access to newspapers, broadcasting and the telephone. Sen himself has gone further by arguing in his analysis of famines, that the existence of a critical free press is an important independent variable determining the occurrence of famine and thus makes a direct contribution to the functioning of both nutrition and mortality. He has made a similar argument in respect of education and especially the education of women showing, in-comparing Kerala and China on the one hand with the rest of India on the other, that it is the variables of education and other indices of female emancipation, that explain variations in life expectancy rather than differences in per capita national income. Kerala, for instance, has life expectancy levels similar to the United States or Western Europe. Thus the existence of the capability that a free press provides may be very basic indeed.

Thinking of entitlements in terms of functionings and capabilities allows us to get behind the superficial indices of access and usage that we so often use. While it remains important for the assessment of comparative advantage or deprivation to have figures on media or telecommunication purchasing and usage patterns, we must always remember that these are very crude indicators and do not get to the heart of the matter. We cannot assume that people derive equal functionings from a given usage or that they are in the position, because of other inhibitors, to take advantage of the full capability set that is in principle on offer. just as Sen argues that people have different capacities to translate a given food bundle into nutrition and also have different nutritional requirements to reach the same level of functioning, so in the field of communication too, it is the real availability of opportunities and the real achievement of functionings that matters. To take a crude example, levels of literacy clearly determine to what extent people can use modes of communication based on print — newspapers or a postal service, just as we know that levels of education not only determine what *people* can get out of their media usage but also what types of media they seek out and use. Crucial for the capabilities approach the evaluation of these differing levels of capability and functioning is something that can in principle be objectively measured and is not, for reasons we have examined, a matter of personal preference.

But what the capability approach highlights is that access is not enough. In evaluating levels of entitlement we need to take into account both the range of communication options made available, and these must be real options and not mere choices between products and services with minimal real differences, and the ability of people actually to make use of these options, to achieve the relevant functionings. We can have real reasoned disagreements as to what range and type of service can now be regarded as a necessary and about how to equalise the level of achieved functionings. But the first crucial point, from the capability perspective, is that this cannot be justified simply in terms of either the metric of exchange that is what people actually buy or the metric of utility that is what people actually enjoy.

Let me now turn to look at communication policy from this perspective. I will start with the mass media and then turn to telecommunications. We need to think of newspapers and broadcasting as enablers of a range of functionings rather than as a stream of content to be consumed. We can then judge the media and communication policy on the basis of how well or badly they serve these needs and how the relevant capabilities are in fact socially distributed. Thus newspapers and broadcasting contribute to a state of belonging to a given cultural group or society. One is not, in a sense, and feels oneself not to be, a normal citizen if one is unaware of the major stories in the papers and has not had the opportunity to view and listen to a range of commonly viewed and watched radio and TV programmes. These are the currency of daily life. At this level the issue of freely available vs subscription clearly arises and the fact that the move of popular sporting events from free over air to subscription, at least in Europe, has raised such widespread public opposition, is an index of this functioning.

But the capabilities approach enables us to take a much finer-grain approach to the problem and to distinguish between different potential uses of the media and the uses actually made and to ask why potentialities available are not actualised (it is in a sense Uses and Gratifications revisited but without the gratification). Thus the media may contribute to quite distinct functionings. A programme may provide information on politics and thus contribute to the functioning of political participation. It may provide information on health care and thus contribute to the functioning of health. It may provide information on the job market or on consumer products. It may provide knowledge and experience of cultural practices. It may provide education in both the broad and narrow senses. The point from a capabilities perspective is the assessment of what contribution the medium makes to enhancing its users' range of functionings and thus to their optimal level of being and doing. We might thus from this perspective advocate the public service provision of broadcasting on the grounds that market provision fails to maximise the capability set of broadcasting users.

But this is only, important as that is, to look at the problem from the supply side. We also need to know what the users themselves do with the opportunities presented and to what extent they do not achieve the maximal state of well-being — to the extent that they fail to use the programming to achieve the functioning of informed political participation or improved health care or better choice of job or participation in a wider range of cultural practices, where the barriers lie. These inhibitors may be crudely economic, to be addressed by simple economic redistribution, but we also know that other, perhaps more intractable ones, lie deep in the structures of social stratification and within what Bourdieu calls the habitus. They clearly and importantly relate to education. But the point from a capability perspective is that if we are serious about entitlement in this field these inhibitors need to be addressed and with positive discrimination. What it does not allow us to do is to use the cop-out, common both within the media and elsewhere, that the media are the way they are because they are giving people what they want.

Let us now turn to look at telecommunications from this perspective. Welfare entitlement in telecommunications has been traditionally expressed in terms of universal service. Let us leave to one side the fact that most of the claims made historically for universal service, are mythic. How does the capabilities approach change our view of what universal service might entail. What should the relevant entitlements be? At present, they are defined largely in terms of crude access, i.e. dial tone supplemented increasingly by various lifeline and low-tariff services. We need to think of entitlements to telephony in terms of a range of functionings. There is, for instance, the being in a state of contactability. We can argue, 1 think, from the Smithian perspective of what produces social shame, that in developed societies at least, the ability to say ring me or I'll ring you without thinking about it, has become the social norm and that those without that potential functioning are disadvantaged. But the absence of this ability also affects other important functionings. It is an inhibitor of the maintenance of familial and wider social networks and thus of full participation in the social life of the community. It may also be a powerful barrier to full participation in the labour market. From this perspective, those who have access to telephony through public call boxes, lifeline service etc. are clearly disadvantaged. The existing established functionings of telephony are not being made universally available or to put it another way, the telephonic capabilities of this proportion of the population are severely restricted and this will in turn contribute to narrowing the capability set across a wider range of possible functionings — employment, participation in cultural events, dealing with public bureaucracies, access to health care etc. As in the associated area of transport, its lack or limits clearly narrows the range of options for movement through social space.

This approach then raises the question of when new options should be added to the capability set as a matter of entitlement. For instance, while I personally hate the mobile phone, we have to recognise that mobility widens the capability set and is a functioning that is increasingly widely available. Are those without it now sufficiently disadvantaged to make it an entitlement? Similar questions are already being raised in relation to Internet connections, usage of the World Wide Web, etc.

I do not pretend that these are easy questions. But what the capability approach argues is the need to develop and agree more finely grained measures and indices that reflect what people in practice can or cannot do with these services, the benefits they do or do not derive from them, rather than measuring mere access or expenditure. We can then devise distributional policies to establish, depending on our chosen value scheme, either a common base level or equality of capability.

Let me then conclude by summing up the argument. The welfare state is concerned with entitlements. This then raises the question, entitlements to what? In answering this question Sen developed the capabilities approach. This rejects both utilities and resources as the appropriate space for the evaluation of entitlement and shifts to functionings, which measures well-being in terms of being and doing, and to capabilities, defined as the range of options of being and doing a person has available. The rejection of utility as the appropriate measure of well-being and the advocacy of a needs-based rather than either desire or merit-based approach to entitlement means that in principle capabilities and functionings can be evaluated objectively and comparatively through a process of rational, public debate based upon the Kantian shift to the position of the other. It moves away both from a post-modern relativism of value and the view of the consumer as free at the point of consumption and from the neo-classical, welfarist view of the free market as producing the maximisation of individual utility towards a more Aristotelian view of objective value based upon basic human functions and therefore needs. In evaluating the performance of the media and telecommunications and of communications policy from a welfare perspective, the capabilities approach moves us away from the metrics of both money and pleasure towards the ways of being and doing enabled by communication and towards an analysis of the barriers that stand in the way of people actualising the social potential communication both do and could make available. It also leads to policies of positive discrimination in the face of these barriers if we are serious about equality.

Yves Poullet

Liberties and the Information Society: the Right to Participate in the Information Society and the Right not to

Our intention is to underline how the development of the information society should lead to affirming the existence of two new rights: the right to participate in the information society; the right not to. These two rights should be considered as complementary, they appear as the very preconditions for the survival of our liberties within an information society that wishes to consider itself democratic. The content of these two rights can be inscribed under different headings as we will show below.

1. The right to participate in the information society

It is frequently stated that the development of information and communications technologies contributes to democracy, notably by allowing us to more actively take part in society, inasmuch as technology offers everyone a means of public expression and favours the activities of associations whose objective it is to take an active part in the public debate.

However, the myth of the global village invoked by the partisans of the above affirmation risks to remain just a myth if certain precautionary measures are not taken. If we are to avoid a two track society, the right of every citizen to take part in the information society implies that each and everyone has to have access, at a reasonable price, to the technical means necessary to be present in cyberspace, as well as access to certain digital information. To this first right a second is added. The citizen who is admitted to exercise his freedom of speech within the information society, must be able to master that which arises from the cyber-image he/she creates by exercising that freedom. This right to control the use of this electronic identity constitutes the second part of our deliberations.

1.1 Right of access, or how to redefine the universal service within the information society

The idea of an evolutive, universal access service to communication technology was first preached in the USA, then in Europe. If universal service is designed according to official directives and documents, whether European or American, as a service which gives qualitative accessibility to all at a reasonable price, its range should be progressively enlarged according to the evolution of demand. Though today limited solely to voice transport and access to an infrastructure of limited capacity, the concept could tomorrow include the availability of E-mail, navigator programmes and access to an electronic signature, all at an easily affordable rate. Maybe the laws of the market as put forward by Pareto's optimum, will render any state intervention in this matter useless. If however the development of these media by governments and the private sector were to expand to such an extent that not being connected could risk depriving the citizen of a number of vital advantages, such an intervention would become justified tomorrow.

The right of access we are concerned with here cannot be seen as purely individual. As we see it, the demands of democratic debate require that numerous associations of a philanthropic, cultural, religious or philosophic nature should be able to use these technologies in order to promote their ideas and contribute to the level of public debate. The idea of making training in the use of modern communication technologies available to associations for the design and maintenance of their sites, even of financial support for such projects, was recently raised. We know the precedent set by 'Community TV', managed by associations whose goals are similar to those cited above. Some of these associations have been specially trained in the making of material for TV within a state-subsidised scheme that makes the transmission of their programmes obligatory.

The access to technology will only become reality if, in parallel, the right of everyone to receive adequate training in the use of new media is recognised. The struggle against technological illiteracy, recently described by the French as 'illectronism' appears to be a second aspect of the universal service. Recent measures taken by the Belgian government as well as at the regional level to connect educational establishments to the Internet and to set up a mixed fund to this end, half public — half private, aiming to provide schools with the neccesary equipment and programs for students to realise this wish. We should add that this right to education, or rather to modern means of education, justifies the appeal for a complete exception to the prerogatives of copyright and associated rights for the purposes of research and education.

If individual and collective rights of access, both to technology and to education will prove to be indispensable prerequisites for a citizen's participation in the information society, access to certain informational content would appear to be even more fundamental if such participation is to be assured. In this context it has been noted that the requirements of functional transparency imposed on governments, in particular by laws concerning access to administrative documents, have not been conceived in a purely passive sense, which is to say that the state is obliged to respond to the needs of the citizen. They are seen in an active sense, as the state's obligation, after consideration of the imperatives of privacy, state security and public safety, to make the information it holds available to its citizens, and this by electronic means whenever feasible. This affirmation, intended as a recognition of a genuine universal public information service, is already law in several countries. Thus Australia and the United States have already adopted the 'Electronic Freedom of Information Acts' obliging administrations to make any general interest information they possess available on the Internet free of charge.

The duty of the state does not, however, end there; the rule of proportionality has been invoked to affirm that the state must also make the same information processing tools available to its citizens which it uses itself to process data. Thus a citizen could acquire a program, used by the administration for the calculation of tax or social assistance grants, at a reasonable price. A recent law on administrative simplification preaches the same gospel, calling for the creation of universal access turnstiles for small and medium-sized businesses, to enable them to simplify their relationships with the authorities and to grant them access to public sector data. This idea could be extended to individual citizens who would then be able, via a single electronic turnstile, to contact their administrative authorities, talk their case through with them and fill in the required electronic forms. The right to electronic procedures would thereby be recognised.

The universal service concept, when applied to informational content, does not only target the state. First American jurisprudence, but more recently the Flemish Legislature, have invoked the duty of cable system operators not to restrict public access to certain information and current events judged to be in the general interest. The case in question concerned television transmissions of sports and other current events which diffusion could not be encrypted. Is not the obligation of the private sector to make information that serves the general interest available to the public already present in law, such as in environmental legislation requiring the publication of pollution or radioactivity levels?

Recently, American doctrine focussed on the possible emergence of a new role for the state with regard to a society where the source of journalistic information is increasingly private and risks being guided by partisan bias or economic interest. The idea, and it is an astonishing one to have surfaced in the land of unbridled liberalism, was to make the state responsible for creating a body to guarantee that information is neutral and free from all pressure. In our own case this consideration justifies maintaining a universal (or public) journalistic information service capable of serving as a point of reference for the entire population.

1.2. The right to 'control' one's own computer image in the information society

The internaut leaves multiple traces in many places in cyberspace each time he or she enters it, whether through the use of search engines, such as Altavista, when surfing from site to site, when taking part in one or other of the available discussion forums, buying the latest best-seller on Amazon, making a flight reservation at a favourite travel agent, or when sending a message to a friend, a member of parliament or the family doctor.

It must be made clear that the internaut may be involved in processes other than those he or she has legitimately chosen, invisible processes such as 'cookies', programmes like Java Script which have been activated by servers in response to questions about the user's net navigator, invisible hyperlinks inserted by various parties, such as cybermarketing services, into the websites he or she consults in order to build up a user profile of each surfer's choices.

To these risks can be added others which are linked to the very nature of the Internet. The global character of the net, being ignorant of all borders, allows data processing activities to take place in localities which offer no protection. The 'open' character of the network not only makes it vulnerable and multiplies the risk of data confidentiality being breached but may, above all, prevent any control of the finalities to which data are used. The information I make accessible via the web, my response to a particular statement made within the forum of a discussion group, for example, may be tracked down by search engines — and Deja News has made a speciality of this — and used in an entirely different context to an entirely different end.

The risk of increasing opacity in data processing and the risks of uncontrolled proliferation must be countered by an enhanced and increased affirmation of the right of the person concerned to their transparency. Given the limitless and borderless nature of cyberspace, it is all the more essential that the right of each and everyone to know who is putting them on file, how, why and exactly for whom, be recognised therein, just as it already is in the traditional world.

Here the principle of proportionality, as treated above, finds new applications with regard to relations between governments and the governed. An automatic harvesting of data by those responsible for its processing must be matched by their obligation to ensure transparency by electronic means. Thus the right to access one's own data, to insist on its suppression or correction or to oppose its transmission to a third party must be automatically possible by means of a single on-line consultation.

Doubtless such a demand will not elicit a satisfactory response unless and until it leads to concrete measures aimed at ensuring respect of our national law outside these frontiers in countries where agencies collecting or processing data are sited but where such rights are either non-existent or flouted. To this end the necessity for instruments of law emerges, effectively enshrining the above-mentioned rights and ensuring their respect, to be established in every part of the globe.

This is the sole condition capable of assuring citizens against the risk of expressing themselves personally on the net. Numerous studies, notably in the United States, demonstrate that the lack of adequate data protection is felt by citizens to be obstacle number one to their actual use of information super-highways. It is clear that insuring adequate data protection is the evident guarantee of individual free expression. It is therefore impossible to affirm the one without the other.

2. The right to exclude oneself from the information society, or towards a new right, that of 'opacity'

The first section sought to clarify those legal instruments capable of assuring a citizen's participation in cyberspace, the second introduces a flat note to the harmony of participation thus encouraged. This is the issue of recognising the right not to participate, that is to say, the right not to appear as a participant on the information super-highway (right to anonymity), even to refuse the use of technologies of information or communication. Beyond this we raise questions on the limits to be imposed upon the use of technologies; whether we should not be reaffirming the individual's right to privacy, not in the sense already mentioned, that of control of one's cyber-image, a control which ensures a better dialogue between the citizen and the authority or company administering his or her data, but rather in the primary sense of the term 'privacy': the right to be 'left in peace', that which we have chosen to call the right to opacity.

2.1 The right to anonymity

Not long ago, on 3 December 1997, a working group calling itself the 'Article 29 Group', created by the data protection directive, concluded as follows: 'The ability to choose to remain anonymous is essential if individuals want to preserve the same protection for their privacy on-line as they currently enjoy off-line...'. The right to anonymity, the recommendation makes clear, is based as much on the right to privacy as it is on the right to freedom of expression and demands that the legal restrictions justified by important issues of public interest should be strictly proportional to the necessity for the maintenance of such interests within a democratic society. This right to anonymity is expressed by the right to make use of anonymous methods of payment and encryption technology. It extends to the right not to reveal one's own telephone number in an advanced telephone network and to the right to use means of anonymous access (public Internet terminals or a de-personalising server).

2.2. The right not to make use of new information and communication technologies

This second aspect of the right of citizens to exclude themselves from the information society may firstly be considered as founded upon social and psychological conditions: thus, how can we expect an elderly person, suspicious of technological progress, to use a communication system that he does not understand? There is, however, a more essential significance. The freedom of each and every one of us is expressed in our choice of communication method. There can be no question, under whatever pretext of efficiency or administrative or budgetary consideration, of imposing on those seeking to obtain access to a service or see to some formality, the exclusive use of an electronic system. The duty of authorities to maintain, alongside electronic procedures, traditional paper methods, without in any way penalising those who ask to use them, seems to me in need of affirmation. Thus a person's right not to use an electronic signature is justified.

2.3 The right to be left in peace, or the right to opacity

A recent resolution of the International Office of Work called for workers to have access, on the one hand, to spaces not covered by video surveillance, and on the other, to communication media (telephone, e-mail) through which the absolute confidentiality of messages, vis-à-vis their employer, could be assured. The resolution was justified by the steep increase in technologically assisted surveillance of behaviour of the workers.

In an entirely different field, data protection and consumer protection directives have enshrined a new right: that of opposing, simply upon demand and free of charge, any electronic processing of one's data to commercial ends. The right not to appear in telephone directories enshrined in the 'telecommunication and privacy' directive is derived from the same principle.

That this amounts to the first burgeoning of the right to be left in peace is indisputable. However, this right to opacity looms larger and more fundamental than that. Some countries have forbidden the processing of data relative to a viewer's programme choice by cable television operators, and the use of data relating to the hiring and purchasing of videocassettes is equally subject to regulations.

The low cost of data processing, the almost infinite transmission and processing capacity, explain the temptation for private or public operators to use information technologies to increase the efficiency of their systems, enable easier control and combat abuse. Thus we already find ourselves, and shall increasingly in the future, taking part in the setting up of vast computerised systems, inaugurated, either within the social security system, to ensure optimum administrative and financial management of the health-care sector, or in other sectors, to enable fiscal control or control of the black market in employment. In urban areas, there is the temptation to multiply the number of automatic monitoring systems to resolve the difficulties of peak-hour traffic control or to combat car theft. Illegal traffic in telecommunications, including Internet crime, will become the processing target of various operators in the name of public necessity: this will be a matter of providing the police with the appropriate arms to combat cyber-crime, child pornography and other excesses to which the Internet has become a home.

Faced with the infinite possibilities for the use of information and communication technologies, the duty exists to abstain from the proliferation of processes and connections, so that citizens may freely exercise their right to self-determination within the secret garden of their own private space.

Conclusions

The information society demands the recognition of new rights of which the first signs already emerge through certain texts, though sparse or of limited range. There is a programmatic interest in recognising such new rights. They enable us to put their first appearances into perspective and to plead for their enlargement and enhancement in proportion to the rate at which their context evolves. The affirmation of these new rights clearly points towards two values considered to be essential in our society, but which risk hypertrophy in the current climate of new technologies: rights of property (including intellectual copyright) and the public interest.

Thus technology (for example Electronic Rights Management Systems) risks granting to copyright, the existence and principles of which are incontestable, a dimension and effectiveness that is without precedent. It must be clear, and we should be reminded, of the fact that these technologies and rights may not prejudice our common right of access to knowledge and that, where necessary, the right to knowledge limits the prerogatives of bearers of such rights.

Thus public interest, which represents the access to knowledge, legitimates the limits of such prerogatives. On the other hand, the same public interest, inasmuch as technologies tend to grant it increasing and sometimes disproportionate effectiveness, must itself be determined by the citizen's right to free space.

So, we find ourselves called upon to reflect profoundly on the equilibrium between the rights of citizens and those rights the power of which, whether in public or private domain, is unquestionably reinforced by technology. This debate is a fundamental one if we are to survive in the face of such powerful agents, not just as isolated individuals, but as citizens who are fully responsible for our social choices.

Bart Cammaerts

Economic and Political Restructuring, Social Citizenship and New Social Rights in the Information Society[1]

This article will point to the necessity to review and enlarge current communication policies. Before doing so this article will focus on the precarious position of the state on two interfluent levels. First the crisis of the sovereign nation-state will be analysed quite extensively. The second level of analysis focuses on the welfare state. Communication and more specific ICT constitute a grow sector in such a way that they appear to have a large impact on both levels of analysis. When speaking of an evolution towards the information society the sovereign welfare state still plays a crucial role. This article will thus conclude by examining new modes of regulation designed to preserve the historical link between communication and public interest.

1. Crisis of the Sovereign Nation State?

The nation state as a sovereign state hedged within a geographical area of material, economic, physical, and psychological independence (Deutsch, 1980: 117) no longer exists. The ever-growing complexity of the international political and economic system makes it nearly impossible for nation states to control every aspect of those systems. In order to cope with this complexity parallel, non-sovereignty-bound structures have emerged. Examples of these structures are trans-national corporations, (I)NGOs, scientific networks, inter(trans-)national organisations, et cetera. Some authors refer in this context to a shift from the traditional Westphalian sovereign state-centred system towards an interdependent multi-centred system (Rosenau, 1990: 252; Rosso, 1995: 179; Hirst & Thomson, 1995: 422). To point this out we must only look at the characteristics of the modern state as it came into being at the end of the 19th century: monopoly on violence, territoriality, sovereignty, bureaucracy, legitimacy, citizenship and taxation (Weber, 1978: 54-56; Pierson, 1996: 8). As you can see in the enumeration below, state power is being undermined on almost all fronts:

- Violence <= defence treaties
- Territoriality <= growing economic and political interdependence
- Sovereignty <= new patterns of governance
- Bureaucracy <= inefficient, impersonal, time-consuming
- Legitimacy <= questioned by anti-state rhetoric
- Citizenship <= individualism, exclusion of large parts of the population (dualism)
- Taxation <= limits set by the market (capital-labour), unemployment, demographic shifts

Indeed, the reason behind this crisis of the nation state cannot be reduced to one clear-cut explanation. According to the so-called globalisation theses the world is shrinking. Metaphors such as the global village or spaceship earth point to a global context where borders become obsolete and values are universalised (Ohmae, 1990; Fukuyama, 1989). There is of course ample evidence that we are witnessing a thorough transformation not only of the world system but also of societies as such. Some authors see globalisation as the propagation of the capitalist world economy on a global scale (Wallerstein, 1984: 18). The transition of a fordist economy to a post-fordist one can also be seen in this context. Others refer to information and communication technologies as being the instigators and catalysts of the globalisation process (Rosenau, 1990: 17; Zacher, 1992: 99). The modernisation of the communication infrastructure is seen as enabling increased capital mobility, a new international division of labour and cost-efficient transport.

The problem however is that popular use of the 'G-term' has blurred the concept up to a point where it becomes an empty catchphrase. What makes the globalisation notion even more problematic is the rhetoric that surrounds it as such globalisation is being projected as a fait accompli: inevitable and value-free (Ferguson, 1992: 73). If we would have to choose keywords to encompass current evolutions, only 'international restructuring' and 'interdependence' come close (Ruigrok & Van Tulder, 1995). Interdependence is a more adequate term than globalisation. It refers more to power relationships and concepts of governance. Current international restructuring can be approached on two levels: an economic and a political one. In terms of international economic restructuring this paper will analyse increased capital mobility, introduction of competition world-wide and changing patterns of industrial strategy. In terms of international political restructuring changing patterns of governance will be highlighted. By doing so I do not want to disregard other factors of importance as pointed out above.

1.1 International Economic Restructuring

1.1.1 Consequences for social policy of increased capital mobility

An important factor in this analysis has to do with the international capital market, the fuel of the economy. Mobility of capital has increased dramatically in the last 15 to 20 years. The average capital flows from 17 highly developed democracies[2] as a percentage of GDP rose from 5% in 1965 to more than 50% in 1993 (Swank, 1998: 673). In 1995 it was estimated that for each dollar in the productive economy, 20 to 50 US dollars circulated in the purely financial economy. Each day more or less a trillion US dollars changed hands. States only accounted for a mere 14 billion US dollars (Korten, 1995: 48). The rest could be considered fluid capital which is managed by actors who have for a large part disengaged themselves from the sovereign nation states. The power they represent is enormous, as pointed out by the crash in Mexico[3] in 1994 and the recent crisis in Southeast Asia. These crises are mainly the result of doubts regarding the growth potentials of the so-called 'emerging markets' in Latin America and South East Asia. The severe socio-economic consequences of their, often short-term, decisions are not included in the daily or quarterly results being presented to the customers of the investment houses. In this regard, Alexander Lamfalussy, former president of the European Monetary Institute, points out that:

> 'The real problem of capital movements and all those financial innovations is that they put a lot of pressure on national policy. (...) The margin for policy has become much smaller.' (De Morgen, 8/11/1997: 70).

The international capital system thus leaves states with very limited room to stipulate a proper financial policy. The only thing they can do is control prices, and thus inflation, and limit the public deficit. If states do not play by these rules, they run the risk of speculation on their currency and of missing investments, essential for employment (Schmidt, 1995: 78). According to the Group of Lisbon the consequence of this logic is a fundamental crisis of the legitimacy of the state which is not compensated by an increase in legitimacy as a result of new forms of world economy control to counter the current liberalisation and deregularisation forces (Groep van Lissabon, 1994: 16). Capital mobility not only weakened the legitimacy of the nation state, but as Geoffrey Underhill observed it 'has led to a restrictive economic environment for those broadly associated with welfare state coalitions, with a redistribution of resources' (Underhill, 1997: 315) (cf. infra).

1.1.2 Dogmas regarding the world-wide introduction of competition

Current discourse dogmatises two basic assumptions; 1) that full competition on all levels is a precondition to economic growth and prosperity for all, and 2) that interventions by public authorities has a restraining, rather than enabling, effect on economic growth. Grazia Ietto-Gillies refers to this when she says:

> 'The underlying assumption (...) is that globalisation leads to increased competition and to the efficient allocation of world resources world wide. This makes governance unnecessary as markets become self-regulating.' (Ietto-Gillies, 1996: 16).

Closely linked to the notion of competition is that of free trade. The reasoning goes that more free trade will create more openness and lead to more choice. More competition on the other hand will lead to cheaper prices, all to the benefit of the consumers. Criticising this idealist-globalist vision proves to be very difficult. How can one be against more choice and lower prices? The question however is whether the present shift towards more free trade and competition will result in positive effects which the proponents use as the donkey's carrot. Especially if we consider the second assumption, namely that of the auto-regulatory capacities of the market. Historic and theoretic evidence suggests the contrary (Underhill, 1991: 198). Full and fair competition rarely exists, it is an idealistic situation. Market actors traditionally aspire to dominate the segment(s) in which they are active. The force behind this mechanism is the maximisation of market share and profit. Concepts like oligopoly, cartels or regulatory capture are strikingly absent from the idealist-globalist analysis. In this context we can refer to the recent wave of mergers in almost all segments of economy, but mainly in the service sector (e.g. banking, insurance and telecommunication). Considering the example of telecommunication, one can conclude that concentration strategies, closely linked to the convergence concept, are designed to dominate production as well as distribution (Melody, 1997a: 198). A world of abundance is being projected, while scarcity is being created. This tends to contradict the assumption that more competition leads automatically to more choice and cheaper prices.

This leads us to the second assumption, namely that the state is a constraining rather than an enabling factor in the economic process. For the first time since the end of World War II, the image of a minimally guided market is seen as the best regulating mechanism (Burgelman, 1996a: 53). At the same time the call for liberalising trade policy, deregulating the economy, privatising the '(x-)inefficient' public enterprises and dismantling the welfare state has gained momentum. The world-wide propagation of liberalisation, privatisation and deregulation by neo-liberal politicians, eminent think-tanks, industrial elites and international entities such as the Worldbank, IMF, OECD, et cetera have severely undermined the legitimacy of the state and its public service (Haque, 1996: 188). As seen above, states have to abide to the new international financial order by controlling their inflation and deficit. The time of deficit spending is over. Growing economic interdependence and the WTO-liberalisation of the service sector make that those services have to be provided in a competitive context. Moreover the modernisation of infrastructure is seen as essential to remain competitive. These facts, combined with the willingness of private capital to do the necessary investments, moved privatisation and liberalisation higher and higher up the political agenda. Privatisation and liberalisation are however not the same as deregulation or freeing the market from constraining state intervention. If considering the liberalisation of telecommunication infrastructure in the EU, one could observe that the privatisation of the PTOs and the opening of the market, led to an inflation in telecom legislation and to the establishment of regulatory bodies to protect public interests inherent to ICT. Thus, liberalisation and privatisation lead towards the necessity to re-regulate. This again contradicts the idealistic assumptions.

1.1.3 Changing patterns of industrial strategy

The globalisation of industrial strategies refers to a shift in the strategies of industry from basic global strategies, multi-domestic strategies or export-based strategies towards complex global strategies (Ducatel, K, 1998). The result of these complex global strategies is a further divergence between industrial interests and national state interests. As Ricardo Petrella puts it:

> 'The national space, the national economy and the national market no longer constitute the beginning and the end of relevant strategies for economic actors, more specific from the private sector (who 'pull the strings' and fix priorities)' (Petrella, 1997: 50).

This coincides with the rising dominance of TNCs over the world economy. Estimates point out that they are responsible for more or less four-fifths of world trade (Ietto-Gillies, 1996: 8) and two-thirds of world production (Third World Guide, 1995: 49). 'Evidence also shows a strong increase in the number of companies that are transnational and in their geographical spread (see figure 1). Raymond Vernon (1979: 258) studied a sample of 316 US and European multinationals. Within that sample the number of large American TNCs with foreign affiliates in less than six countries decreased dramatically from 138 to 9 (76% -> 5%) between 1950 and 1970. The European sample shows the same shift, going from 116 TNCs active in six countries towards 31 (86% -> 23%). The other side of the story is an increase in the number of companies active in more than six countries. Between 1950 and 1970 we see an increase in the US-sample from 43 to 172 (24% -> 95%) and in European sample from 19 to 104 (14% -> 77%). We can easily assume that this trend was consolidated during the 1980s and 1990s.'

Increase in the number of TNC present in >6 countries (1950-1970)

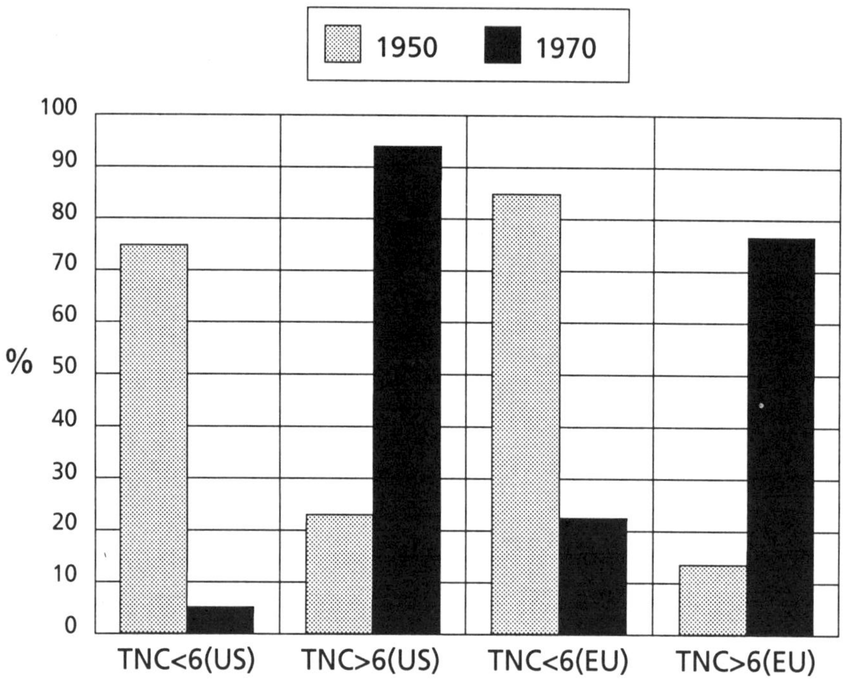

This shift can be explained by the 'international vertical integration'-strategy adopted by TNCs. According to the (macro-micro) economic conditions and the nature of the product, components are located in different countries. Once processed, the component moves to another country where it can be integrated in the production cycle. By doing so, TNCs managed to heighten efficiency, lower wage costs and create a platform for penetrating foreign markets. (Ietto-Gillies, 1996:8) Although not every company followed this strategy we can observe a general tendency towards a new international or more appropriate regional division of labour. Large American and European corporations moved parts of their production to countries where they could find cheaper labour, flexible labour conditions and other cost-cutting advantages like a favourable tax system (Fröbel, Heinrichs, Kreye, 1986: 211). Governments on the other hand often adhere to the classic economic thesis that taxation hinders investment (OECD, 1990). This results in governments competing for investment by lowering or even abolishing taxes, which in turn heightens the pressure on labour as a prime source of state revenue.

1.2 International Political Restructuring

As explained above, international capital and crucial parts of industrial activity have loosened their 'preferential' relationship with the nation state and their 'public interests'. It is therefore only logical that the above-mentioned techno-economic transformations, have also changed international politics quite drastically. The political restructuring consists in a shift from the sovereign state to the interdependent state. In a sense we are witnessing changing patterns of governance. These new modes of governance system can be described as multi- or polycentric, whereby the state constitutes one level of a complex system of overlapping and interacting areas over which states exert only limited control (Rosenau, 1990: 272). The room for states to go solo has indeed become smaller as Mark Zacher clarified:

> 'The key is that states are becoming increasingly enmeshed of collaborative arrangements or regimes that are creating a very different international world (...) from which exit is generally not a feasible option' (Zacher, 1992: 60).

As shown in figure 2, states have to take the position of international political and economic actors into account whom they share an interdependent relation with. This, however, does not imply that the final point of this paradigm-shift will be the end of the nation state. Nation states remain very important players but they are no longer the only ones on the field. A cluster of interrelated factors makes it nearly impossible for the nation state to isolate itself from the international politico-economic game. A game in which they do not solely control the rules anymore.

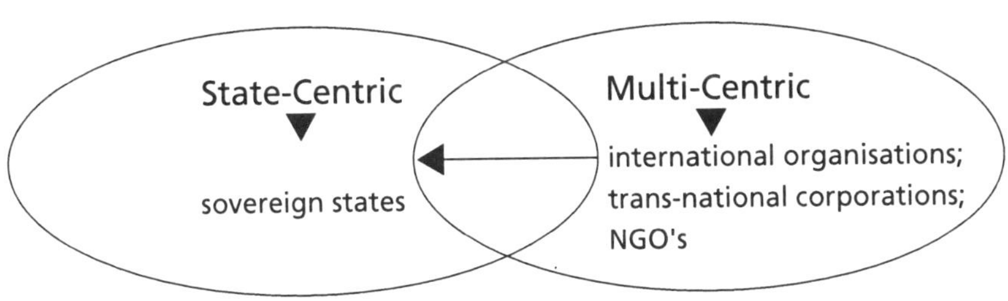

In order to retain a certain amount of control, states have conceded their autonomy regarding economic and financial policymaking to trans (inter-)national and regional entities (e.g. EU, OECD, NAFTA, ASEAN, MERCOSUR, G7, WTO, et cetera). Many of those organisations focus solely on (free) trade, thereby freeing economic actors from national 'social' constraints (Schmidt, 1995: 76). Social and fiscal policy, however, did not follow the same logic. This is often referred to as the 'democratic deficit'. Here, the weakening of the state consists in the increasing difficulty for the state to stipulate a consistent socio-economic policy. The impotence of politicians to 'fully' solve the unemployment crisis in Europe proves this point daily (Petrella, 1994: 16). This, however, does not imply the end of the nation state. Earlier, the relationship between the state level and the others was described as being interdependent. They need each other in order to attain their often divergent goals. Furthermore, nation states are constitutionally legitimate powers. In the light of TNC dominance this constitutes an increasingly important asset. Nation states may have encountered setbacks when it comes to sovereignty, but they still have a pivotal role in relaying international levels of governance and an ever-growing articulate population (Hirst & Thomson, 1995: 432).

In this context the appearance of the 'corporate governance' concept in the idealist-globalist rhetoric is not surprising. Authors like Ohmae assert that the world economy is being driven by two main factors; global market forces and TNCs. For idealist-globalist theorists the *new* global system will be governed by market logic, degrading public policy to a secondary role (Ohmae, 1990). The corporate governance — concept also assumes that corporations embody public interest values. In this logic the citizen becomes synonymous with the consumer. As the term 'idealist' already confirms, these visions are far from realistic. First of all, they are disturbingly undemocratic. Secondly, one could argue that the governing powers of the nation-states have diminished, but that does not automatically mean that they have vanished. State and market relate to each other in interdependence. As Hirst and Thomson point out;

> 'Markets and companies cannot exist without the protection of the public power. The open international economy depends ultimately on Western (particularly US) force and upon active public regulation backed by legal enforcement' (Hirst & Thomson, 1995: 427).

More and more we can also observe new emerging mechanisms of governance which challenge the often top-down approaches in the analysis of state power. These evolving forms of governance also refute the rigid public/private distinction (Jessop, 1995: 310). In the 'civil society'-ambit e.g. (I)NGOs are sometimes successful in denouncing actions by governments or TNCs. In a sense they carry out a 'globalisation from below' through global networks of solidarity (Kellner, 1997: 10). These civil organisations use ICTs as a complementary strategy to exert influence. In this respect we can refer to:

- the world-wide campaign to save Mumia Abu-Jamal[4], an American political activist, from death-row
- the Greenpeace media campaign against the intentions of Shell to dump the oilrig Brent Spar in the North Sea[5]
- the Global March Against Child Labour and the 'Clean Clothes'-campaign[6], focusing on the precarious conditions of (child)workers active in developing countries for big corporations such as Nike, Levi-Strauss, H&M, Reebok, et cetera
- the recent coalition of labour unions, non-governmental development agencies and even some governments against the MAI[7], proposed by the OECD. This resulted in new negotiations which started in October 1998.

These actions show that counter-forces can successfully oppose strategies or decisions by governments, TNCs, or powerful international bodies like the OECD. Besides using ICTs, (I)NGOs also develop novel strategies to exert influence. Greenpeace, for instance, buys small amounts of stock in order to get inside the shareholders' meeting of the company they want to criticise. For instance, the French-Belgian energy holding Tractebel was criticised by its minority shareholder Greenpeace for its plans to build a pipeline in a valuable piece of rainforest, home to a vulnerable indigenous population (De Morgen, 26/05/1998). Such counter-forces will have to be strengthened and further developed. If we focus on the (tele)communication sector, one can observe a striking absence of consumer organisations in current discourse. They do exist, only we rarely hear them.

2. Crisis of the Welfare State?

New theoretic insights (cf. Keynes, Beveridge, Marshall[8]), the economic boom of the 50-60s and the relative success of social and liberal democracy in Europe, made that the state expanded its economic activities exponentially. The welfare state of the 60s positioned itself midway the rigid communist state and the harsh capitalist state. Authors often refer to socialised capitalism alluding to the solidarity mechanism inherent to the welfare state. The free market is acceptable only when it can be controlled by society and is directed towards the development of a socially just society (Petrella, 1997: 34; Witte, 1990: 73).

2.1 The welfare state as a broad political concept

The corner stone of the welfare state can be described as social citizenship, a concept introduced by T. H. Marshall in his classic work '*Citizenship and Social Class*' (Marshall, 1950). Other authors have refined this rather vague concept. Interesting in this respect are the analyses of Gøsta Esping-Andersen. Besides formulating different ideal types of welfare state regimes, he explicitly linked social citizenship to the granting of social rights. In a sense Esping-Andersen equates social rights to property rights, inviolable and granted on the basis of citizenship rather than merit. In this way social rights can be viewed as decommodifying the individual vis-à-vis the market (Esping-Andersen, 1990: 21). A classic example here is the so-called 'social contract' between collectivity and citizen. The contract embodied a certain number of rights:

1) right to employment
2) protection against poverty
3) protection against social risks
4) promotion of equal chances (Burgelman, 1996: 53).

The welfare state as a broad political concept was, of course, much more than the granting body of social rights. It was also marked by a compromise between capital and labour in order to make capitalism democratically acceptable as the dominant system for organising society. The basic feature of that compromise was gradual and even growth of consumption and production. Thus social citizenship and the welfare state also refer to different regimes of social stratification and income redistribution (Esping-Andersen, 1990: 58).

Another characteristic of what has become the welfare state are public services. As Andrew Calabrese points out:

> 'Welfare policies sometimes can and do represent a collective value placed upon state support for what are understood to be basic pre-conditions for effective participation in social, political and cultural life, and for the cultivation of civic competence more generally' (Calabrese, 1997: 2).

In this regard, provisions such as education, promotion of art and culture, public libraries, public broadcasting, universal access to telephone and energy should be considered part of the welfare state as well. Public services are regarded as being in the public interest and as such states have generally protected these services from market forces. Historically speaking, public services moved from the private sphere, churches or charity, to the state and its institutions, although in some countries a strong system of subsidiarity remained in place (Esping-Andersen, 1990: 61).

Three kinds of public services can be determined. The first kind could be described as legal services. These refer to security, defence and the justice systems. Furthermore, a distinction is made between public services which also have economic goals besides public interest ones and public services with a non-economic mission. The former refer to services such as gas, electricity, post or telecommunication, the latter to services as healthcare or education;

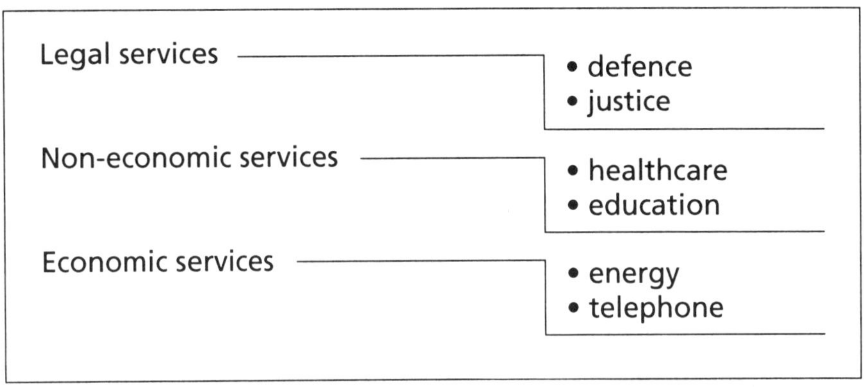

Figure 3

This distinction has the advantage that it is clear-cut. The disadvantage however is that it becomes easier to disregard the public interest values also inherent to economic public services.

2.2 The welfare state under pressure

Many authors have emphasised the influence of economic and political restructuring on every aspect of the welfare state (Giddens, 1994: 136; Petrella, 1994: 62; Pierson, 1991; Burgelman, 1999). Full employment is no longer a sustainable goal, equal chance provisions and national solidarity are put under pressure by an individualistic utilitarian consumerism model and the system of social insurance is undermined by budget constraints and demographic shifts. Furthermore we can observe a considerable weakening in the negotiating power of the labour movement when active in social talks or struggles. As has already been said concerning public services, a clear distinction was made between the 'economic' services and the 'non-economic services'. The latter being provided by the state, the former conceded to a regulated market. The reasons behind the liberalisation and privatisation have been explained above, but in the light of the welfare state, we have to keep in mind that services such as telephone, energy, mobility, et cetera are more than mere 'products'.

Some critics see in this evolution the repression of the welfare state paradigm by a new one which accepts poverty, unemployment and other social evils as being 'natural' (Zsuza, 1996: 6). In this regard Philip Cerny speaks of the 'Competition State' and Bob Jessop introduces the 'Schumpeterian Workfare State' (Cerny, 1990; Jessop, 1994). Within these new paradigms the state is no longer an enabler but a disturbing factor to economic growth and free trade. The best regulatory mechanism is considered to be a minimally regulated market. Similarly the minimal state is put forward as an alternative for the, lazy-making, welfare state (Hayek, 1960). This, however, is misleading. The dichotomy between state and market included in the concept of the minimal state equates liberalisation with deregulation and laissez-faire. This again contradicts history. As Karl Polanyi quite rightly pointed out:

> 'There was nothing natural about laissez-faire; free markets could never have come into being merely by allowing things to take their course (...) Laissez-faire itself was enforced by the state' (Polanyi, 1944: 139).

Surely, the days of the welfare state marked by direct interventionism lie behind us, but the end of the welfare state in the sense of 'social justice' or 'social citizenship' is nowhere in sight. The state, as guarantor for the welfare of its citizens, still has an irreplaceable role to play, be it in a different way than 'in the old days'. This is why the above-mentioned analyses of paradigm shifts have to be approached with some reserve. Even hard-line thatcherism was not able to dismantle the welfare state (Pierson, 1994). In the Worldbank communiqué 'Rethinking the State', Joseph Stiglitz, its head-economist, confirmed that blind faith in the market and a retreat of the state is not the right answer to today's problems;

> 'We now see that markets and governments are complementary: the state is essential for putting in place the appropriate institutional foundations for markets' (Worldbank, 1997).

The OECD also made a modest U-turn when her president Donald J. Johnston stated:

> 'There is general agreement, as we move into the 21st century, that environmental and social considerations are key elements or constraints for future economic development in a globalising world, and thus essential to ensuring the longer-term welfare of our populations' (OECD, 1998).

Many actors are thus becoming aware of the vast consequences of dichotomous state/market thinking for social cohesion and the notion of solidarity within a given society. The consumerisation of society which regards people as customers instead of citizens, has also its effect on the social fabric of a society. Furthermore, it has dawned that welfare is one of the conditions for economic growth. Ken Ducatel, Gustavo Fahrenkrog and James Gavigan, echo this view by arguing that social development is a precondition for sustainable economic growth in Europe. Social development, they say, 'is not just "nice to do", we have to do it' (Ducatel, et al, 1997). The focus has therefore shifted from the welfare state to the question of the means to attain these social welfare goals. Consensus is growing concerning the revaluation of the welfare state concept in a way that transcends the classic dichotomies state/market or economic policy/social policy. The role of social sciences here should be to develop new models and theories which embody this. This however will not be easy, especially because it remains hard to disengage oneself from the normative and ideological debate that surrounds the welfare state and its future.

3. ICT and the need for a new public (tele)communication policy

Between the lines of this paper, information and communication technologies have never been far-off. Indeed ICTs touch upon almost every aspect we have discussed, often on different levels, but nonetheless. For instance, when discussing the increased mobility of capital, implicitly ICTs play their role. They also have their part in the expansion of TNCs through more flexibility and more efficient logistics. Concerning citizenship and the welfare state, we can also observe a growing importance of ICTs in job destruction/creation, in social security or in developing new ways to bridge the gap between the state and its citizens.

The third part of this article will thus look into the public values linked to ICTs and related to a competitive market-oriented context. It becomes apparent that the potential benefits of ICTs can only materialise when on the one hand the welfare/nation states adopt a new (international) logic of regulation and on the other pursue a pro-active policy.

3.1 Regulation as a safeguard to the public interest

As previously concluded, the state still has an essential role to fulfil and welfare remains its basic goal. The means however are to be adapted. The liberalisation and privatisation of public services, such as energy, transport and communication, prompt questions about the public interest values inherent to those services. The 'public interest' is however an essentially contested notion.

In other words it can encompass a lot of meanings. Think of the politician who uses the catchphrase; 'it's in the general public's interest' to stifle opposition or to advantage a dominant actor (Stigler, 1971). Therefore, it is essential to determine how we can translate this volatile notion in the new competitive context and in the interests of the small users, both professional and residential. As states are no longer players in the direct sense of the word, they try to retain control indirectly. In this sense regulation is generally seen as the most adequate instrument to attain social objectives. In scientific terms this notion is a bit problematic. In French this is exemplified by the difference between *réguler* and *réglementer*. The former referring to a broader approach of the regulation notion, the latter to a rather restricted approach.

When it comes to ICTs and telecommunication, regulation refers to the restricted 'réglementer' approach. The most common example of the regulation logic in telecommunication policy is the originally American concept of universal service. A private company is allowed to provide a 'public' good, in this case telephony, under a given set of conditions designed to preserve the public interests which are attached to that good. Here again we see that those conditions and the extent to what is regarded to be in the public interest, varies from country to country. A second argument which makes regulation, and universal service in particular, problematic is that historically speaking, regulation has often been (ab)used to protect vested interests (Jessop, 1990; Melody, 1997b: 195). In this respect regulatory capture is a well-documented phenomenon.

William Melody responded to this danger by formulating a 'long-term market regulation' concept. He recognised that competition can provide significant benefits, but is inadequate to achieve economic and social (telecom) objectives. Therefore, he argued, a more independent form of regulation is essential (Melody, 1997b: 198). Regulation also has to adapt to the global/regional tendencies. National regulation has to be complemented with regulation on a regional and a global scale. If the nation welfare state is to continue playing a role of any significance, it must introduce the political process with its democratic values in the context of the international political economy (Cerny, 1991: 194). This said, we still have to ask ourselves whether regulation alone will do the job to attain the social welfare goals in order to achieve sustainable development? From what follows it will become clear that the answer to this question is negative.

3.2 Access & Protection: two key rights to be guaranteed?

As argued above, ICTs are more than mere products. The growing importance of ICTs in every aspect of life affects the rights of citizens as defined in the social contract. Considering the above-mentioned importance of welfare for economic growth, but also the necessity to evaluate and revalidate the current welfare state, one could argue in favour of a realistic revision of classic rights and the extension towards new rights. On the one hand some classic rights such as full employment have become difficult to guarantee in a post-industrial context whilst on the other hand new needs call for attention. Conservatism on the left as well as the right side of the political spectrum has made revaluation and/or adaptation difficult or almost impossible. A way out of this status quo could be the forging of a new social contract based upon consensus and the social citizenship concept.

With regard to social welfare and ICTs, one can distinguish two main concerns which might be eligible for becoming a given right for each citizen, namely access and protection.

3.2.1 Access to

More and more often, access to information and information networks is becoming a pre-condition for participation in economy and in society. As Bish Sanyal and Donald Schön observe, policymakers often ignore the possibility that the information society can bypass low-income communities (Sanyal & Schön, 1999: 377). They quite rightly state that:

> 'We, as a society and as individuals, can choose how we shall think and act in relation to the digital revolution. But one thing is clear: to profit from the potentials opened up by IT — whatever they may be — we must participate in it. This is especially true for the poor, who are already excluded from the economic, social and cultural mainstream. They must be helped to participate in the digital world in a variety of ways' (Sanyal & Schön, 1999: 375).

Translated to the above-mentioned social contract, this would imply that citizens are entitled to claim their right to communicate. In this sense, one could argue that universal access would be more appropriate than universal service. An approach to ICT-services which goes further than purely telecom policy seems to be more sensible in this regard. Such a cross-sectorial approach should place access on different policy agendas: not only telecoms, but also social affairs, education, welfare policies and economic affairs. Thus, access here not only means physical access to infrastructure, hardware and software, but also means providing your citizens with transparent information about policy on all levels of government. Furthermore, access also refers to provisions for equal chances, to an adapted education system and adequate possibilities for training. This last issue will prove to be of great importance. Access is not to be understood as mere physical access to an IT-network, but also as acquiring the necessary skills in order to select and assess information. This can also be placed in a 'lifelong learning' perspective (Mansell & Wehn, 1998: 66). As such it would be unfair to ask the market to fully pay for universal access as is the case with universal service. Mixed financing systems which go beyond the regulation logic are to be designed. This however does not preclude a substantial contribution by the private sector as state budgets remain tight. As history tells us, this should be part of a bargaining process between the public and the private sphere. To be successful, the private sector has to be persuaded that it is also in their interest to have more welfare and a better-educated work force.

'We are entering an era of unprecedented abundance', is an often-heard slogan. In a technological sense, this may be true but it remains to be seen whether abundance will be able to close the ever-growing gap between the information-privileged and information-impoverished (Schiller, 1986: 102). Recent evidence shows that, as Manuel Castells puts it, 'the informational city is a dual city'. In the article with the same name, Castells warns against the dangers of what he calls a 'systemic urban schizophrenia', but also projects the prospect for a better future: more humane, more intelligent and more just (Castells, 1999: 28). By doing so he will be criticised for being idealistic, but the case can be reversed for so-called pragmatists in the sense that they accept and cultivate the status-quo of social and other forms of exclusion.

3.2.2 Protection from

If we on the other hand consider the implications of ICTs for privacy and the dim prospects of real and fair competition, a second right emerges, namely the right to protection. Regarding privacy matters, the stakes are very high. Personal data have become a tradable commodity. The cookie-technology has introduced coercive surveillance on the Internet. On a long-term basis these practices lead to an erosion of trust in people's minds (Samarajiva, 1997: 432). This lack of trust is not only damaging in economic terms, but also in political terms. It relates to the already weakened relationship of trust between citizens and the state. Trust in society and the state is however a dynamic concept which requires a long-term pro-active policy directed at restoring confidence in the state and its institutions (Giddens, 1990: 121; Schoechle, 1995: 446). Besides privacy, another issue calls for attention. There is considerable reason to fear the emergence of dominant actors in the (tele)communication market. Accepting competition in the telecommunication sector could lead towards an unacceptable concentration of power, especially in reference to the convergence of information, telecommunication and computer (Mansell, 1993: 9; Kalkman, Hins & Jurgens, 1995: 6). The goal of liberalisation cannot possibly be the substitution of a state monopoly by a private oligopoly. Competition, therefore, has to go hand in hand with anti-trust policies. It is essential that pluralism of opinion, diversity of supply and fair competition is enforced upon the market. Protection in relation to ICTs can also be viewed in other fields. For instance, the protection of consumers/citizens against different forms of abuse from operators or service-providers. The awareness concerning the necessity for developing ways to assure a sound protection is not only growing in political circles but also within industry. The different auto-regulatory efforts point in this direction. The illicit content debate is another story. There is a broad consensus that child-pornography is to be prosecuted. Beyond that it becomes very blurred. What about racist/nazi propaganda? What about the use of ICTs by international crime organisations? These questions clash with the 'freedom of information' logic of the Internet and with data-protection concerns. This often moral debate falls however beyond the scope of this paper.

3.2.3 Sector-Specific Provisions and Cross-Sectorial Provisions

Within these two areas, access and protection, we can make a further distinction between those provisions which can be regarded as sector-specific (namely telecom) and provisions which go beyond the telecom sector and can be labelled as cross-sectorial (cf. Tab.1).

1) Access, seen in a sector-specific perspective, refers to universal service, interconnection and standardisation as the guarantee for affordability and quality of service. Important here are modes of regulation directed at, and in favour of residential users. Access, seen in a cross-sectorial perspective, refers to a policy directed at avoiding dualism and promoting equal opportunities for participation. Regarding 'equal opportunities' we can think of the right to information and transparency of policy. 'Avoiding dualism' refers to social regulation, such as barriers to disconnection, public freenets, price reductions or media education.

2 Protection, seen in a sector-specific perspective, refers to anti-trust policies and consumer protection. The latter can be viewed in terms of transparent and sound procedures for complaints, independent mandates for consumer commissioners, et cetera. The former in terms of avoiding abuse of market power by incumbent operators, policing merger, et cetera. Protec-

tion can also be approached in a cross-sectorial manner. In this respect the case of data protection and illicit content was evoked. Considering data protection, it is important to keep in mind that legislation is not the only aspect that has to be looked at. The enforcement of the law and the means of tracking abuses should also be taken into account.

4. A classification of the roles of the state in the Information Society

When we bring together all the elements put forward above, we can make the following classification:

	ACCES	PROTECTION
SECTOR-SPECIFIC	• universal service • interconnection • standardisation	• anti-trust policies • consumer protection
CROSS-SECTORAL	• universal acces • right to information	• data protection • illicit content regulation

Tab. 1

I use this classification for the sake of analysis. This however does not imply that the different, defined roles of the state in the information society are exclusive to the category to which they are designated. For example, one could argue that an anti-trust policy also applies to a cross-sectorial approach. Information is crucial to a citizen in a democratic society. When that information is provided by a very limited number of sources, the scope within which the citizen observes the world becomes limited as well. The same goes for universal service. Some countries have included in their universal service obligations some very distinct social concerns leaning more towards universal access than towards service. This refers to measures such as free calling time for people on welfare or limits on the possibilities to disconnect.

In a sense, the classification tries to conciliate both economic and social goals. In some cases this is very difficult as they tend to diverge. This however is not always necessarily true. Partnerships between industry and the state, directed at different but well-defined social goals, should be encouraged. As Bob Jessop observes:

> 'Although public money and law would still be important in underpinning their operation, other resources (such as private money, knowledge or expertise) would also be critical to their success. In this sense the state's involvement would tend to be rather less hierarchical, less centralised and less dirigiste in character' (Jessop, 1995: 324).

But industry will also have to adapt itself and recognise its social obligations towards societ(ies)y.

5. Conclusion

It has become apparent that the widening of the scope towards other policy areas such as social policy, education policy, economic policy and fiscal policy will be essential in the development towards the information society. Although this sounds logical, it often is not. In Belgium, for instance, this widening proves to be very difficult due to a complex constitutional context, whereby the federal government is responsible for telecommunication and fiscal policy and regional governments are responsible for media, social matters and education[9].

The welfare/nation states must develop a two-track policy approach. The first track is an economic- and market-oriented regulatory framework axed on the communication sector. The second track is a more social and mixed regulatory framework axed on a multitude of sectors such as education, social affairs, equal chances, fiscality, et cetera. A mixed framework refers to a balance between state intervention (through subsidies from the social budget for instance) and social obligations to be paid by the industry (through a system of funds for instance). The effective implementation of such a dual framework, will however not only depend on the willingness of the industry, legislation and enforcement by state institutions such as regulatory agencies and anti-trust bodies. Consumer organisations and other social actors such as labour organisations and consumer commissioners should exert constant pressure on the industry and government agencies when representing the users/citizens. In a sense, they must forge a counterforce to corporate dominance and build a dyke against regulatory capture. Only in these circumstances can we hope that competition in the field of ICT services will bring genuine benefits to the public at large.

It would be fair to say that we are a long way off from such an approach, but when contemplating regulation, supervision and socially-inspired m(i)acro-economic policies, it is difficult to disengage oneself of normative choice. After all, we are concerned with questions regarding the kind of (global) society we want to live in.

Notes

[1] This article is the result of an ongoing interuniversitary project 'Research & interdisciplinary evaluation of the information society: networks, usage and role of the state', funded by the Federal Institute for Science Policy (DWTC). Partners are CRID-CITA (Fundp), LENTIC (Ulg) and SMIT (VUB).

[2] Australia, Austria, Belgium, Canada, Denmark, Finland, France, West Germany, Ireland, Italy, Japan, the Netherlands, Norway, Sweden, Switzerland and the United Kingdom.

[3] 52 billion dollars in loans, mainly from the US and the IMF, were necessary to stabilise the peso and thus the Mexican economy (press reports, December 1994)

[4] see http://www.xs4all.nl/~tank/spg-l/mumia002.htm or http://www.mumia.org/

[5] see http://www.greenpeace.org/~comms/brent/brent.html
versus http://www.shellexpro.brentspar.com/

[6] see http://www.cleanclothes.org/1/

[7] The 'Multilateral Agreement on Investments' gives extensive rights to foreign investors (mainly TNCs) without any duties attached to those rights; see http://www.xs4all.nl/~ceo/mai/ or http://www.islandnet.com/~ncfs/maisite/

[8] For a detailed analysis of these authors see: George, V. & Page, R. (eds.): (1995) 'Modern Thinkers on Welfare', PrenticeHall/Harvester Weatsheaf, Hertfordshire

[9] For more on this see Lobet-Maris, C., van Batselaer, B.: (1996) 'Belgium, a Laboratory for Europe: The Federal Model and the Information Society', *Telematics & Informatics*, Vol 13, nos 2/3; p. 81 and the article of Claire Lobet, Béatrice van Bastelaer and Bart Cammaerts further in this book.

Claire Lobet-Maris, Béatrice van Bastelaer & Bart Cammaerts

On the role of Government in the Information Society

Introduction

The current agenda of liberalisation of the telecommunication and media services markets raises important questions about which role government has to play or wants to play in the development of the so-called information society.

This question is of course not specific to the telecommunication and media sectors, but is related to a more general concern concerning the role of the state in economic development. As pointed out by J. E. Stiglitz (1996):

'The dramatic failure of the grand socialist experiment had an unanticipated consequence: it lent support to extremists of the opposite ideological persuasion, according to whom government should play almost no economic role. But the rejection of one extreme is not the affirmation of the other. The real issue that both the success of East Asia and the failure of the socialist experiment raise is, what is the appropriate role of government in the economic development? There is (...) many third ways between the extremes of total government control of the economy and complete laissez-faire' (Stiglitz, 1996: 11-12).

In his paper, Stiglitz also argues that '*government definitely has a place but it must know its place*' (Stiglitz, 1996: 17).

Concerning information society initiatives, national or regional governments often adhere to different scenarios regarding their roles and the types of action they endorse.

This article is based upon different national studies elaborated in the framework of a TSER-DG12-project called *Social Learning In Multimedia* (SLIM). This study analysed the cases of Belgium, Denmark, France, Germany, The Netherlands, Norway and United Kingdom.

This study focused primarily on policy output rather than on policy outcome. This relevant distinction in policy analysis was put forward by Volker Schneider (1997) in a recent article. For Schneider, policy output concerns public action programmes which address policy goals whereas policy outcome is considered to be '*the transformation of the programme into reality*' (Schneider, 1997: 344). In the field of the information society, policy output as presented in the different SLIM national reports are for instance national action plans, governments' declarations, programmes of actions and trials, regulatory texts and actions.

The central question of this article could be: do politics matter? This question will be answered by comparing national plans and policies. So the answer will remain somewhat theoretical to a certain extent since an adequate answer to this question should be based on facts and realisations rather than on plans and programmes.

1. In Search of an Analysis Framework

Analysis of the policy output of a nation or a region raises two main questions:

1. why do public policies intervene in the field of the information society?
2. how do public authorities define their responsibilities or their roles in this matter?

The first question is concerned with the goals of, or the motives for, public action whereas the second relates more to the patterns of public authority action.

1.1. Goals of Public Intervention: Market versus Society

Theoretically speaking, the action of public authorities should be motivated by the public or general interest. Regarding the information society development, public authorities often use quite different definitions of what is in the public interest. The translation of this contested notion varies from one country to another, which in turn influences the goals and actions endorsed by the respective public policies. The public interest can be motivated by economic goals and issues and thus oriented towards the development and competitiveness of the national communication and information industry. In that case, public interest is perceived to be above all market or industrial interests. But public authorities can also focus more on societal goals. In that sense, technological and industrial developments are not ends in themselves but only means to achieve certain social or political goals deemed to be in the interest of society as a whole.

An indicative element in the way public authorities define their goals regards the position of the information society issue into the division of ministerial responsibilities. In most cases we see that the information society issue resorts under the political responsibility of the Ministry of Industry or Commerce. This expresses a quite different political orientation than when it is managed by a minister who also has cultural responsibilities.

1.2. The Pattern of Public Authority Action: Interventionism vs Laissez-faire

The second dimension to be considered concerns the degree of government interventionism. As pointed out by Stiglitz, there is a continuous range of possible political positions between two extremes that are on the one hand the absolute 'laissez-faire' and on the other rigid 'control' or 'interventionism'. The pattern of public action is related to the way public authority defines its responsibilities towards society. In the 'laissez-faire' perspective, as argued by Schneider, 'market is treated as the natural state of society and the "mother of all problem-solving" ' (Schneider, 1997: 342). In that sense, the intervention of the state in a matter as the information society is generally moderate and oriented towards the correction of 'market failures'. In contrast to this economic perspective, the interventionism approach considers the state as an essential actor, having a crucial role and responsibility in the transformation and evolution of society. In this regard, it is the state that is considered to be the 'mother of all problem-solving' and its intervention can refer to economic issues as well as social concerns.

1.3. Domains of Action

A third level of consideration when comparing national policies concerns the domains of public actions. Through comparing US, EU and UK policies, William Dutton et al. (1994) have identified four major domains of state's action in the ICT field:

- *'a guardian of the public interest*
- *innovator in its own use of the technology and catalyst in encouraging non-profit activities of benefit to different sectors of the public*
- *regulator and referee to create a market environment that encourages economic growth and technological innovation*
- *general cheerleader in stimulating the take-up of ICTs'* (Dutton, 1994: 35).

These four domains can each be detailed by way of specific actions a government can endorse.

1.3.1. Regulating Actions

Those actions can be subdivided into two categories of regulation: market regulation and societal regulation

a. Market Regulation
In a context of full liberalisation, the role of the government as a regulator appears to become more and more important in order to ensure fair competition between operators and service providers. This regulation role can be ensured through different actions:

- enforcing fair competition
- speeding up standardisation
- extending intellectual property laws
- negotiating equitable conditions of access to the competitive market at world level.

b. Societal Regulation
Societal regulation concerns specific actions a government should take to avoid the development of a dual information society and to guarantee the public interest. This societal regulation concerns mainly

- the definition of universal services and provisions policy
- the protection of privacy
- the security of network systems.

Some of those regulatory roles can be endorsed at a national stage, others need international consensus negotiated on a European or world-wide level.

1.3.2. Promoting Actions

This second set of actions concerns the role the government can play in order to stimulate the use of multimedia. This 'promoting' role can be sustained by different mechanisms (such as subsidies, financial/fiscal advantages, etc.) and should be oriented towards different user groups. Four main components of promoting actions can be identified:

- promoting use and application in a residential context
- promoting use and application in public administrations
- promoting use and application in the non-profit sector
- promoting use and application in the profit sector.

1.3.2. Educational and Innovatory Role

This third set of actions concerns the overall role a government has to play in encouraging and organising the acquisition of basic knowledge to use multimedia and exploit its potentials. This concerns not only education policy, but also the stimulation of training with regard to life-long learning.

2. Analysis of National Policies

This section will be devoted to the analysis of information society policies endorsed in each concerned country. This analysis will mostly be based on the SLIM reports1. We will only describe those policies from an internal perspective. The comparative aspect will be dealt with in the next section.

2.1. UK Policy: a Liberal Policy Framework (Steward & Slack, 1999)

Stewart and Slack argue that government's action in the UK is concerned mainly with the promotion of competition in all fields of industry based upon regulation rather than large-scale public investment. In that context, according to Stewart and Slack, the government regards national multimedia capabilities and infrastructures as a priority. However, competition in the telecommunication sector, and to a lesser extent in broadcasting, have tended to restrict the establishment of a coherent policy for the UK information society that goes beyond the level of statements of intent (Stewart & Slack, 1999).

The principal statement of UK policy is set out in a 1994 document:
'The government remains committed to facilitating the continuing development of the full range of communication by:

- most importantly, establishing and maintaining a stable, effective and evolutionary regulatory framework
- working with the private sector to articulate a clear vision of the place communication can play in the UK's national , economic, social life
- promoting the competitiveness of UK operators, manufacturers and suppliers of the full range of telecommunications and services
- setting an example by being innovative in the government's own purchase and use of telecommunications and services
- giving due consideration to the needs of rural areas for new communication services, and continuing the commitment to universal service in telecommunications

- obtaining maximum benefit from UK government and European research funds available to support communication development
- keeping abreast, and taking account, of the development of new technologies in communications
- giving due consideration to the protection of personal and corporate data on networks
- providing appropriate protection to intellectual property rights.'(Board of Trade, 1994)

The role of the government in the UK appears to be limited to industrial objectives, and more specifically in terms of fostering competition. Moreover, policy seems entirely oriented towards telecommunication's infrastructure and services concerns. UK policy can therefore be regarded as a liberal one, where that market is regarded to be the main driving force for the information society development. In that sense, the role and responsibility of the UK government are restricted to the correction of market failures and the intervention as a regulator of market competition.

The UK policy paradigm is thus based on the 'free market'-model as the source of innovation and development. As put forward by Stewart and Slack (1999), the laissez faire stance of the UK towards multimedia resulted in intervention being limited to regulation, as opposed to specific investment in large scale technology projects. As a consequence of this laissez faire policy, the UK lacks a coherent vision on its multimedia industry and the direction it should take. Instead an amalgam of views are differentiated by the interests of the respective major players in the UK market.

2.2. France: the Pragmatism

Thierry Vedel (1997) and Béatrice van Bastelaer (1999) have developed a very comprehensive view on the French policy in the field of the information society.

For Vedel:

'the French state has long demonstrated an inclination toward high-tech ventures and a propensity to launch grants projects. Yet this kind of state entrepreneurship often termed ' high tech colbertism', may be vanishing with the information society' (Vedel, 1997: 307).

Compared with its previous history in support of telematics, cable TV or high-definition TV, France's first steps in the information society domain appear to be rather modest, at least under the rightist government. The government has limited its role to accompanying private initiatives. In this sense, French information society policy may indicate a smooth, though radical shift in government-industry relations, from dirigism (administrative guidance) to pragmatism.

This shift can be explained by many factors. First of all, the pragmatic stance endorsed by the French government could be explained by the recent failures of large public policy projects in the high-tech domains, such as the 'cable plan' termed 'le syndrôme du plan câble'. A second factor of importance in this new orientation of French policy regards the liberal views of the government and the lack of public resources to launch ambitious policies. Thirdly, Vedel points towards the reluctance of France Telecom (the national telecom operator which in the past was a major player in the French public high-tech policy) to play a central and instrumental role in the information society. Finally, according to Cédric Curtil (1996), this restrictive role of the French government could be explained by the fact that telematics in France is already a mature market which does not need a strong intervention of the public authorities anymore.

In sharp contrast to this restricted political perspective, two reports commissioned by the French government in 1994, stressed the importance of Information technology for French society. Both reports point towards the major role the French government should play in this domain. The first report, issued in June 1994, was prepared by Thierry Breton (1994). It underlines the importance of new teleservices for the French economy and employment. In this report, Breton argues that the French state has a major role to play in supporting the development of new teleservices, and especially in the field of health and education. The second report was commissioned to Gérard Théry (1994) and issued in September 1994. It designs the main lines of a French action plan regarding information society. As underlined by Vedel, this plan is based upon three proposals:

- 'All French households and firms should be connected to the IS* by 2015
- The IS necessitates a massive development of fibre-optic links
- Experiments should be initiated immediately' (Vedel, 1997: 316).

Compared to the ambition of these objectives, the French Plan for the information society, as defined by the governmental council in October 1994, appears very modest. In that plan, the French government designs its role in very restricted terms and limits itself to two main actions:

- 'Adaptation of the regulatory framework: the aim is to allow more competition while maintaining basic principles such as universal service, the protection of privacy and social cohesion
- Stimulation of new markets and services: public administration is expected to launch applications that will create a virtuous circle within which the IS can develop' (Vedel, 1997: 318).

Summarising the French policy, Vedel concludes:

> 'Overall, French IS policy is mainly reactive — even defensive — driven by economic and industrial considerations. As with previous high tech developments, the main issue is not what people need and want and the social utility of new infrastructure but the position of the French industry on the world market' (Vedel, 1997: 310).

2.3. Belgium: a Federal and Industrial policy (Lobet-Maris, et.al, 1999)

It is hard to talk in general terms of the Belgian policy. In fact, as a federal state, Belgium has no unique policy in the field of the information society since IS policies are designed at different levels of government. Even if regulation of the telecommunication market is under the responsibility of the federal government, the Flemish and Walloon regions have progressively taken the lead in the formulation of information society policies. Often represented as competing policies, the Flemish and Walloon policies share the same rationality. Both of them are strongly marked by an industrial policy approach in which the intervention of the state is motivated by market failures and lack of private investment. This policy has led to the adaptation of the cable network in order to provide new (internet) and/or old (telephone) services. In Flanders, Telenet was endorsed, exploiting the cable as an alternative network for telephony and internet services and in Wallonia the setting up of a regional Intranet based on fibre-optic links called WIN (Wallonie Intranet) was favoured. In the Flemish region, this motivation is reinforced by regionalist tendencies aiming to develop an autonomous Flemish economy whereas in the Walloon region, state investment is supposed to compensate the lack of private initiatives. Even if their motivations are fairly different, the political paradigm that sustains regional investments remains the same, at least at the beginning. It is highly influenced by technological determinism which considers that infrastructure will in itself have a positive impact on the growth of new information services and uses. This regional competition has consequences on the tension between the different communities. Moreover, as competencies are divided, it becomes very difficult to formulate a coherent policy as this requires dialogue. On the other hand, all is not negative. For instance, cheap tariffs for schools and libraries were put forward by the federal government and subsequently implemented by the regional governments.

2.4. The Dutch Policy: the Consensus Approach (Van Lente, 1999)

The orientation of the Dutch policy also appears to be motivated mainly by competition issues and by the typical Dutch tradition of infrastructure building, echoed by the slogan 'Holland – Distributie Land'. In comparison with UK policy, the Dutch government seems to be more than a regulator. Apart from trying to get market players together, it fosters and stimulates new initiatives. Its role appears to be the one of an 'orchestra conductor ' of the information highway and multimedia shaping in the Netherlands. This role is deeply rooted in the Dutch policy tradition based on negotiations and consensus building between the vested interests in the Dutch society. Another Dutch tradition concerns tolerance, ethics and societal issues.

Those two traditions send the Dutch policy in a sort of continuous search of balance between competition interests and societal interests. But in fact competition interests seem to be dominating the Dutch scene of multimedia even if this tendency appears to be compensated by the attention for social and ethical issues:

> 'The aim of leading position for the Netherlands within Europe is an ambitious one. It can only be realised through a major effort on the part of all concerned. The proposed liberalisation of the electronic network, telecom and media services markets, and the adjustment of framework regulations, will create scope for innovation in industry. Continuation of the debate of the public domain will provide greater insight into government responsibilities in this area' (Dutch National Action Plan, 1994, Part II-4. Financing of the Programme).

Another trend in Dutch policy is the recognition of the limited role a government can play in the shaping of the information society. This recognition is not motivated by a liberal obedience of the Dutch policy but by a sort of sociological belief as stated in the 1994 National Action Plan (NAP). Under the heading of 'the government role', the government writes that it *'does not intend to produce a blueprint for the exact design and scope of the information superhighway (...) Only society, in all its diversity, can continually decide which way developments will go'* (Dutch National Action Plan, 1994, Part I — The government's role). This limited role can also be explained by the presence in the Netherlands of large IT&T companies such as Philips, IBM, AT&T and Ericsson, a relatively strong publishing sector and a very active audio-visual sector.

In the Dutch National Action Plan, the government defines its main streams of action:

- 'ensuring broad access to communication media and a wide variety of heterogeneous sources of information for society as a whole (...)
- providing scope for private initiatives for investment in networks and services, through a progressive approach involving liberalisation and, where possible, deregulation of markets for telecommunication and electronic media
- creating a new framework for (self-)regulation, in order to set clear legal conditions for electronic information. Intellectual property rights and privacy are among the issues that play a role here
- organising the government's position as a major user of information (systems) and telecommunication services to ensure that these exert a stimulating and guiding influence on the development of information superhighways

- strengthening the R&D base
- setting up a limited number of demonstration projects, serving as guidelines in the private and public sectors, in order to make a start and gain experience with the development of information superhighways' (Dutch National Action Plan, 1994, Part I — The government's role).

2.5. German Policy: the Industrial Players' Influence (Kubicek, 1999a+b)

The industrial complex of Germany appears to be dominated by a small number of multimedia actors. On the infrastructure side, the incumbent German telecom operator calls the shots. Deutsche Telekom owns about 110.000 Km of fibre-optic lines. Deutsche Telekom also purchased a share of about 17% in the SES Satellite Company, thus becoming the largest investor in ASTRA. Moreover, Deutsche Telekom claims to own and operate the largest cable TV network in the world. In order to protect its tremendous investment in cable and satellite, Deutsche Telekom attempts to expand existing systems through digital compression. Deutsche Telekom is clearly betting on the successful introduction of digital TV. The telecom and cable operator also controls technologies for terrestrial strategy and claims to be very advanced in ISDN. Finally, the German operator has put more optical fibre into the ground than any other telecom company. The network could be used for a 'Highway-strategy' along the American model, but so far Deutsche Telekom has shown little interest. According to experts, this monopolistic situation gives few opportunities to new entrants. It is expected that only two national providers could survive the competition with Deutsche Telekom.

In Germany, the multimedia development scenarios are increasingly influenced by the technological and economic concerns related to the concept of Digital TV. This situation can be explained by the presence of two strong competing alliances around Bertelsmann and Kirch Groups, both providing digital TV, each developing its own standard of set-top boxes, similar in functions but incompatible in norms.

Online services are also dominated by a small number of competitors with Deutsche Telekom and its T-On Line system (0.9 million subscribers) on the one hand, and CompuServe (3.8 millions subscribers) and Bertelsmann (allied with America On Line and expecting to reach 3.5 millions by the year 2000) on the other. This situation might even change towards an even more concentrated market with new alliances to be concluded between Deutsche Telekom and Bertelsmann if the antitrust authority approves of the alliance.

The presence of strong industrial players, highly concentrated, counter-balances the relative weakness of the political actors in Germany. This weakness is embodied into the federal constitution of Germany that scatters the political competencies in multimedia between 16 states or Länder on the one hand and the federal government on the other.

2.6. Norway: a Welfare Tradition (Brosveet and Sørensen, 1999)

Norway is characterised by one of the highest residential Internet penetration rates in Europe. Another characteristic of the multimedia development in Norway is that they seem to have made a choice in favour of telecom on line services, i.e. Internet, and against the development of interactive TV.

The high rate of Internet penetration in Norway cannot be explained by the presence of a strong IT industry pushing the users-market, since Norway has no strong IT sector. With respect to the high rate, two reasons can be given. Firstly, the topography of the country which is favourable to telecommunication solutions and mobile telephony to connect isolated parts of the country. Secondly, the economy of Norway which is very healthy due to the oil-exporting sector which in turn explains the solidity of public finance and the very low unemployment rate (5.2%). This solid economic situation thus allows the government to push some policies in domains such as multimedia, more than is the case in any other European country.

As explained by Brosveet and Sørensen:

> 'while the general economic situation of Norway clearly facilitates the uptake of multimedia, its industrial structure is not particularly favourable. The industrial basis of multimedia is weak, and the industrial structure is not particularly conductive to the widespread use of this kind of technology' (Brosveet & Sørensen, 1999).

Brosveet and Sørensen argue that the political attitude of the government towards multimedia is a bit ambiguous and can be approached as a mix between noticeable liberal flavour spiced with serious efforts of deregulation and the perpetuation of the welfare state.

They indicate that 'the main components of technology policy in Norway can be classified under headings such as general education, public support for R&D, and large investment in public infrastructure'.

2.7. Denmark: the National Values Approach (Jaeger & Hansen, 1999)

Denmark's development process of multimedia is very close to the Norwegian case, but with a stronger emphasis on the role of the state in the protection of social values and the welfare of the population. As explained by Jaeger and Hansen:

> 'Even tough the Danish IT policy is subordinated to the EU policy, the Danish government have formulated a national IT strategy, where the public sector is supposed to have a leading role (...). The Danish strategy for reaching the information society is different from most other countries because it regards the public sector as a driving force in the development. It is also different from many other countries because it stresses that the technology have to be a tool for all citizens — also the weak ones, and that multimedia shall be a way to strengthen democracy at all levels'.

This position is well-reflected in the Danish report on the information society made in 1994 in the wake of the so-called Bangemann Report (1994):

> 'The public sector shall be actively involved with the private one and be the leading force in the efficient use of information technology (...) The strategy shall be based on a Danish model where market forces should not be the only forces involved. The strategy must secure some special values:
> • it shall secure free access to information and interchange information
> • it shall support democracy and individual access to influence
> • it shall open the public sector, make it more transparent and facilitate the delivery of better service
> • it shall be used to support the weak ones of the society
> • it shall strengthen the international competitiveness of Danish companies as the basis of the Danish welfare system' (Dybkjaer, L. and Christensen, 1994).

The importance of the public sector as the driving force in the development of the information society contrasts sharply with the prevailing view that the market should be the driving force in the shaping of the information revolution. A strong public sector compensates for the relative weakness of the private sector. Apart from the national telecom operator — Tele Denmark — , no private actor is large enough to play a key role in the Danish multimedia landscape. Tele Denmark is one of the largest telecom operators in the EU. Tele Denmark also aspires to be one of the largest cable TV operators in Denmark and seeks to control the satellite distribution of TV channels to households.

As far as technological choices are concerned, Denmark has opted for the Internet as the privileged carrier of electronically mediated public services. As explained in the 1994 Danish plan:

> 'The Internet is rapidly becoming the most important medium for electronic communications. More and more public administrations, companies and private users are using the Internet for electronic mail and for the retrieval and presentation of information' (Dybkjaer, L. and Christensen, 1994).

3. National Contexts and Political Styles

In section two, we have shown that political actions and plans in order to shape the information society are different from one national context to another. This proves to some extent that there is still space left for a distinct policy in this matter.

In the present section, we will try to explain those differences by pointing out the contextual factors that shape the different political regimes and policies regarding the information society.

3.1. The Contextual Factors

According to Gourevitch (1986), variation in domestic political response to international and techno-economic devices stems from four contextual constituencies.

3.1.1. Production profile

Hulsink (1996) defines this first factor as:

> 'All the tangible and intangible resources, capabilities and endowments of one economy compared to other economies and the way these domestic assets contribute to economic performance and give the economy a competitive edge' (Hulsink, 1996: 61).

As far as multimedia and information economy are concerned, the cross-national analysis shows a clear distinction between a group of countries with a high profile and a group with a low profile.

In the former group, countries such as Germany, the UK, the Netherlands and France can be found. They all have national champions that strongly influence the political line of their government. This influence is twofold. On the one hand, the presence of a national champion diminishes the pressure on the government to respond to international competition. On the other hand, it also reduces political freedom by focusing political interventions and regulations on matters that primarily interest the industrial lobby.

The latter group with a low profile concerns countries such as Belgium, Denmark and Norway. The national economy of these countries is characterised by small-scale industrial structures made up of SMEs that force the government to design a very pro-active and interventionist industrial policy in order to defend the national basis of the economy. Regarding the outcomes of these policies in the field of multimedia, some questions can be raised about the side-effects of such policies on the innovation dynamics and capabilities of the industrial actors.

3.1.2. Organisation of Interest Intermediation

As pointed out by Gourevitch, the system of interest intermediation is the second factor of importance having an effect and influence on domestic policy. Corporate actors, trade unions and consumers participate in policy-making through their umbrella associations which, besides providing certain services to their members, also function as an information and lobbying channel to pursue their policy interests in the policy-making process. On this point too, national context differs from one country to another. Political analysts distinguish two forms of interest mediation systems. The first one is pluralism. According to Lehmbruch (1982, 1984), pluralism is characterised by the predominance of pressure group politics and the lobbying of government agencies and parliament by fragmenting and competing interest groups. The corporatist system, in contrast, stems from the effective participation of organised interests groups in the policy-making process. In the first institutional arrangement, the key decisions are taken or at least strongly influenced by dominant industrial actors, limiting the discretionary control of government. The UK and to a lesser degree France, are well representative of this type of institutional system. As shown by the UK case, this system leads to incoherent and multiple visions concerning multimedia, all related to the major players on the UK market. The corporatist regime is characterised by institutionalised participation of organised interest groups into the policy-making process. Scandinavian countries, but also Belgium and the Netherlands are strongly influenced by this corporatist tradition in 'consensus' policy-making. In those countries, negotiation and consensus between competing interests are at the edge of the political system. Theoretically, this leads towards a more unified and stable political vision and programme. In between stands German policy. Though belonging to the corporatist tradition in its multimedia policy, Germany is strongly influenced by corporate interests coming from some strong industrial key players.

3.1.3. Structure of the State

The way the state is structured constitutes a third factor of influence to a nation's particular response in the field of the information society. Structure of the state refers to the organisation of the political and legal-administrative system. Concerning the information society, one of the structural elements that does in fact influence national responses is the way political responsibilities are divided into different levels of state authority: national, regional and local. On this matter the situation also varies from country to country. The main distinction can be made between centralised states as opposed to federal ones. Countries like the UK, the Netherlands, Norway and France belong to the former group. In those countries, the central state is responsible for the majority of political decisions that are applied by regional and local authorities. Centralised arrangements have the advantage of cohesion and uniqueness of response but are also frequently criticised with regard to a lack of attention for regional or local assets and needs. The latter group of countries are characterised by decentralised arrangements. They recognise the necessity to delegate specific responsibilities to regions or to local authorities. This is the case in countries like Germany, Belgium and Denmark. The advantage of their system is that policy addresses the needs of regional and local communities and can adapt to changes more quickly. The major criticism concerns the fragmentation of power and the difficulty to set up a coherent policy, especially in new political domains such as information society. The result is often that responsibilities are scattered and this leads to subsequent competition between different federal entities. This is particularly the case in Belgium and Germany where the regulatory instruments are divided among different levels of power, leading to responsibility conflicts and the slowing down of the policymaking process.

3.1.4. Ideology and Culture

This last factor is much more intangible and concerns values, norms and blueprints for the socio-economic ordering of society, providing, as pointed out by Hulsink, *'economic motives and social beliefs for societal actors'* (Hulsink, 1996: 61). Political analysis has been very active on this topic, sometimes referred to as the invisible hand of policy. Major ideal types point out a divergence between what we could call according to Dyson and Wills (1983) and Lodge and Vogel (1987) the *'communitarian state-oriented ideology and the individualist state-less ideology'*. The first type sees the state as having major responsibilities with regard to society, being the guarantee for social values like equity, social rights, public interest, collective goods and community needs.

Within the individualist state-less ideological framework, the state retains a limited role and leaves the responsibilities for the orientation of society to the civil society and the bargaining between competing interest groups. The state thus fulfils a regulatory role in order to give equal opportunities to each individual and to foster fair competition between various interests. Information society policies do not escape from these ideal types. This is the case for the UK whose information society policy is marked by the individualist state-less ideology. On the opposite side, a country such as Denmark bases its policy on a highly communitarian state-oriented ideology. In between, the other countries can be placed more towards the communitarian side of the continuum than towards the individualist one, even if slight differences can be found regarding the affirmation of national social values. On this last point, the policies of the Netherlands and Norway appear to be highly motivated by social values. They seek a national consensus with respect to the orientation of society.

COUNTRIES	Production profile	Interest mediation	Structure of the state	Ideology and culture
United Kingdom	High profile	Pluralist	Centralised	Individualist state-less
France	High profile	Pluralist	Centralised	Communitarian state- oriented
Belgium	Low profile	Corporatist	Federal	Communitarian state-oriented
The Netherlands	High profile	Corporatist	Centralised	Communitarian state-oriented
Germany	High profile	Corporatist	Federal	Communitarian state-oriented
Norway	Low profile	Corporatist	Centralised	Highly communitarian state-oriented
Denmark	Low profile	Corporatist	Centralised	Highly communitarian state-oriented

Table 1. Summary of national contexts and policy styles

3.2. Four Patterns of Political Styles

The analysis of information society national policies and contextual factors (see Table 1) clearly shows that divergences between countries exist. In this last section, these differences will be conceptualised by identifying patterns of political styles.

Those patterns can be identified according to the two basic axes we defined at the beginning of this article: the axis of the goals pursued by public actions confronting market vs society and the axis of the pattern of public action opposing laissez-faire vs interventionism.

3.2.1.The Liberal Style

This first pattern is very endemic to the political style of the UK. Theoretically, this style aims at stimulating market forces and reducing the role of government intervention wherever possible. The basic postulate of this style is that the market is the motor of the societal development. The liberal style does however not mean an absence of political intervention. In the UK context, the role of government appears quite strong, but in a limited way. Its role focusses primarily on market stimulation and the setting up of a strong domestic industrial basis. For instance, the UK government has played a very early and pro-active role in liberalising the telecommunication sector. It is also very pro-active in education in order to develop the necessary know-how to sustain the growth of domestic industry. Conversely, the promotion policy developed by the UK government is quite passive; domains of promotion are being informally designed according to industrial interests and market potentialities. Another characteristic of this political style is the strong influence of industrial lobbies on the policy-making process. Since the market is seen as the main problem-solver of societal issues, the government is very receptive to the pressures exercised by industrials in order to warrant their interests. As pointed out before, this approach leads to the inability of the state to set up a coherent national plan for the shaping of the information society.

3.2.2. The Corporate Style

The second pattern concerns two countries: France (before the socialist government) and Germany. Sharing many of the characteristics of the liberal pattern, the policies differ slightly because of the dominance of corporate interests as put forward by domestic industrial champions. In this framework, the policy making process is dominated formally and informally by the vested interests of large domestic industrial corporations. Political actions and plans appear to be much more those of some industrial firms than those of a government acting in the public interest. This pattern is different from the English one. The liberal style deployed by the UK government is more pluralistic and tries to foster true competition between all the industrial players and their interests. The corporate style is much more based on the collision between government and some powerful industrial players who try to dictate national policy. This makes the state more interventionist than in the liberal scheme but this intervention, meaning regulation and/or promotion, is deeply guided by the specific interests of certain industrial lobbies. This policy style can bear some attractive economic results through public support of the competitive consolidation of some national champions. On the other hand, it can also have negative effects on the interests of society at large. As far as private interests are concerned, the corporatist style leads to a strong politicisation of the business and to the setting up of institutionalised barriers for entering certain markets. With regard to the public interest, there seems to be little space left for political vision and actions based upon social values. Public interest achievements appear to be conditioned by particular interests that in turn may lead towards a weak legitimacy and low social credibility of the government. Finally, the great political confidence put in the hands of some industrial players by the government can result in wrong or even irrelevant decisions, as has been the case in France with the *plan câble*.

3.2.3. The Consensus Style

This third pattern can be applied to countries like The Netherlands, Denmark and Norway. The major characteristic of this pattern is that the state positions itself as a conciliator of various interests and as a defender of consensual national values through appropriate policies. The state is not seen as a dominant player defining the major orientations of society but as a mediator between the various interests active in society and as a protector of national consensual values. This gives another orientation to the information society policies. In this framework, the determinant of policy is society as a whole and not one or a few of its components as is the case in the liberal pattern. The policies developed within the consensus style are twofold. On the one hand, the state is very active in setting up fair conditions for industrial competition and competitiveness. On the other hand, the state protects national values and societal interests, and acts in societal domains deemed important according to the national values but difficult to sustain in a purely market-oriented policy. As a consequence, the state develops dynamic policies to promote the use of multimedia in public and non-merchant sectors. One important particularity of this promotion policy is that it is societal-oriented. Technologies are seen as new means to achieve societal goals such as democracy, administrative transparency, equity, etc. In the long run, such policies may lead to negative side-effects on the dynamics of the market, but this is perceived as secondary to the promotion policies developed by those countries. One of the political results of this pattern is the formulation of a coherent political plan which induces that all the national actors follow the same societal orientation.

3.3.4. The Industrial Style

The fourth pattern is representative of the Belgian situation. This pattern is characterised by a strong intervention of the state on the industrial level and a strong postulate that industrial progress automatically leads to social development. In this country, the position adopted by the state is central and compensatory. Considering the irreconcilable interests of highly organised corporatist associations in this country (labour, industry, etc), the state regards itself as the only actor that can define the orientation of society as a whole. Although the mechanisms of negotiation are highly developed, the state places little confidence in the political clear-sightedness of those corporate associations. To some extent, the state uses top-down strategies to force its political interest and visions upon the different actors, using negotiation mechanisms to legitimate its orientation.

Due to the weakness of the national industrial basis but also to a long-life political tradition, the state imputes itself a central role in the industrial sphere in order to compensate the lack of private investment and projects. This leads to a state with a strong industrial policy, defining sectors that need a pro-active public support. In the information society ambit, this political pattern has given rise to wide political investment. On the federal level, the goal was to protect the national operator and on the regional level, it leads to the development of alternative telecommunication infrastructures. The two levels of policy are difficult to conciliate, but both levels intend to consolidate the industrial development of Belgium and/or its regions. Two main consequences of this policy can be foreseen. The first one concerns private investment. To some extent, the deep involvement of the state in the industrial sphere can be seen as having the counter-productive effects on the private investment front. The state taking the place of industrials, assuming the risks that normally reside with market actors, does not encourage and in certain cases even discourages private involvement. This can be explained by the fear of the politicisation and the subsequent uncertainty. The second consequence concerns public interest. In practice, strong political investment in the industrial sphere leaves few resources and space for political actions oriented towards the public interest, amongst others the modernisation of public agencies or education.

Conclusion

To conclude, each of these styles can be positioned on the two axes introduced in the first section of this article in order to visualise the specificity of each of those political styles more clearly. At this stage of the analysis, it is important to mention that this typology of roles is a broad typology that could underestimate the fact that in a particular line of policy, e.g. education, promotion, or R&D, a government can adopt a position quite distant from its national type.

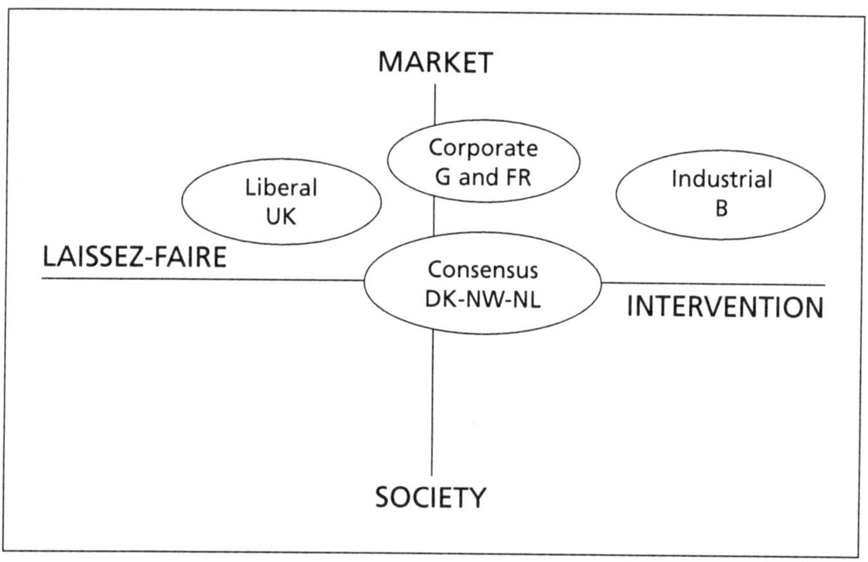

Table 2:
Position of each country regarding laissez-faire/intervention and market/society

At the end of this article, one should clearly keep in mind that our analysis was based on policy outputs and not on policy outcomes. Let us just recall that policy output concerns public action programmes, whereas policy outcome refers to the transformation of the programme into reality. Moreover, the analysis concentrates on multimedia policies pursued during the years 1994 to

1996, missing the changes in policy due to changes in government (mainly in the UK and France).

The main question of this article was: do politics matter? On this matter, national policies show clear divergences in the way national governments act to shape the so-called information society. So politics do matter when it comes to policy outputs. The question remains whether the outcomes and the ability of these policies will lead to different social paths in the way nations construct and appropriate the information society. More research will be necessary to answer this question even if it is the only question that matters, socially speaking.

Notes

1 Even if these reports were published in 1999, most of them were made public in 1996-1997 and concern the years 1994 to 1996. This means, for instance, that for the UK and France, the development of the multimedia in these countries was described before the changing of political majorities and the establishment of the socialist governments.

* IS used for information society

Leo Van Audenhove & Bart Cammaerts

Political Perspectives: Synthesis of the Debate

'Welfare and the Information Society'

The contributions presented in this section—although written from very different perspectives—all highlight social policy aspects of the information society. Taken together, these contributions—and the subsequent discussions at the conference—can be centred around four major research questions:

1. How can social welfare be defined in the information society and how can it be translated into public policy?
2. What are the existing models regarding public interest and are they appropriate in an information society?
3. What are the dominant paradigms used in regulating the information society?
4. What can be learned from an assessment of actual policy and how does it relate to the three questions above?

Within telecommunications policy, the discussion of social welfare (question 1) often tends to concentrate on an enlargement of the concept of universal service. The discussion can be summarised as follows: new services and information in general are becoming increasingly important in every aspect of life as such access to a wide range of services is becoming a prerequisite in order to participate fully in both economy and society. The main question thus is to what extent should the concept of universal service be enlarged to incorporate new services. These discussions are of course valid. However, they rarely go beyond supply-side issues and the technicalities of regulating the telecommunications suppliers to incorporate new services (Graham, 1996: 4). The main objection against an enlargement of the concept of universal service remains that it takes the concept into areas of social policy outside the communications area and loses conceptual clarity during the process. Policies in these areas can not be solely financed by the sector itself and should be covered by other policy concepts and instruments. A second objection is that social policy risks to be reduced to an add-on policy or programme, instead of fundamentally rethinking social policy for a changing society.

The contribution of Nicholas Garnham is refreshing in that, by way of applying Amartya Sen's 'Capabilities' approach to communications, Garnham fundamentally questions our understanding of social welfare and more specifically of public service and universal service. Sen's interpretation of welfare as a 'space of capabilities' — as an alternative set of opportunities a person has, apart from realised functionings — takes the discussion much further than a discussion on available resources or access. It broadens the scope of social policy in that the approach recognises that social, political, economical and individual factors may severely reduce the 'space of capabilities'. These factors are not only part of the communications sphere, but may touch upon very different spheres and may even touch upon the fundamental organisation of society. Policy directed at an equal participation in one sphere may thus require policy actions in other spheres. It should be part of social policy to develop appropriate actions to enhance the 'space of capabilities' for every citizen.

In his contribution, Bart Cammaerts tries to go further by rethinking policy in terms of a right to access and a right to protection, whereby these rights are linked to social citizenship. The concept of right to access should however not solely be restricted to access to technology (or its content). It should also include access to education and skills development. The second concept of a right to protection should encompass concerns about privacy and antitrust. The merit of extending the discussion to protection is that it informs us of the possible negative consequences of information technology, consequences the current 'hype' about the Information Society often tries to minimise.

The article written by Yves Poullet also stresses the emergence of new rights in the information society. Access here is also diversified towards access to technology and education/training, but is seen within the universal service concept which should be enlarged accordingly. Another important observation is the need for more transparency, both from the industry and the state. This touches upon the protection-issue. Quite rightly, Yves Poullet points towards the lack of adequate data protection as the prime obstacle to the widespread use of information technologies. Finally this article raises the very interesting issue of the right to exclude oneself. This refers to the right to anonymity, the right not to use and the right to be left in peace.

However, the contributions by Nicholas Garnham, Bart Cammaerts and Yves Poullet remain predominantly normative. Although all contributions are valuable in appraising concepts of social welfare in communications, they do not elaborate on translating them into actual public policy. In the subsequent discussion Anders Henten remarked that one should refrain from idealising to much about the welfare state. Although the Western state is a social state it remains predominantly capitalist. Every social policy is linked to economic and thus budgetary boundaries. In a capitalist society, social welfare is not only related to redistribution, it is just as much related to the creation of wealth. In this respect, state and market should find ways between redistribution and wealth creation in function of social goals.

The point made by Anders Henten brings us to the contribution by Claire Lobet-Maris, et al.. This contribution elaborates on the third and fourth question in that it analyses the role of government in developing the so-called Information Society. By comparing national government policy in different countries, the authors try to find an answer to the question: do politics matter? The general answer to their question is positive, as different types of state and welfare regimes seem to formulate different answers in terms of policy focus — social vs industrial — and policy means — state vs market. The outcomes depends on contextual factors such as the production profile, the organisation of interest intermediation, the structure of the state, and ideology and culture. Another important finding of the work of Claire Lobet-Maris et al. is that governments seem to believe policy can have a more or less substantial impact on the development of at least some aspects of the Information Society. At first this finding seems to contradict the rather dim picture of the nation state in crisis, as presented by Bart Cammaerts. However, both points are not unreconcilable. What is needed is an understanding of the extent to which processes of international restructuring do constrain the power of nation states and in what areas. In other words, what is the space left for social policy?

We end this summary by referring again to the four questions outlined above. The contributions in this section and the discussions at the conference contributed in elaborating these questions. Much more work however needs to be done. By concentrating on these questions and by basing policy for the Information Society on sound scientific research on past and present policy experiences in the areas of communications, social welfare, economic and industrial policy we may arrive at new insights valuable for future policy, which transcend the current often dichotomic thinking of market vs state in policy formulation (see Burgelman, 1997a).

PART II

Social Regulation

Jan Servaes

Introduction

'Towards A New Media Or Social Policy For The Information Society?'

The policy debate on communication and information has come a long way. Jan Van Cuilenburg and Denis McQuail distinguish between three different historical phases of media policy in the US and Western Europe. During the first phase (until World War II) media policy was largely dominated by the tensions between state and corporate interests at a national level. Afterwards (from the fifties into the eighties) a shift took place from economic and national concerns to more socio-political considerations. This phase is often summarised with reference to public-service broadcasting as the political ideal for media policy, notably in Western Europe. There was a strong policy commitment to universal service, diversity of content, democratic accountability, public financing and non-profit making. Jean-Claude Burgelman argues that these concepts are largely insufficient in view of the problems and challenges that new information and communication technologies pose for the information society. His perspective on the information society is adopted from the definition drafted by the High Level Group of EU-experts: 'The information society is the society currently being put into place, where low-cost information and data storage and transmission technologies are in general use. This generalisation of information and data use is being accompanied by organisational, commercial, social and legal innovations that will profoundly change life both in the world of work and in society generally' (European Commission, 1997a: 11). Such a broad perspective coincides with the third phase, as identified by Van Cuilenburg and McQuail. They describe how from 1980 onwards several technological, economic and socio-cultural trends have fundamentally changed the context of media policy.

The EU industrial policies have changed during the eighties from a defensive towards a more offensive policy. Two sectors where this policy change has become very obvious are telecommunications and informatics. This has led to a technological convergence of communications and computer technology into Information Communication Technologies (ICT). This convergence will have considerable implications for policy formulations at distinct levels. However, it is feared that the EU is not really anticipating an overall policy on the problems of convergence within the EU. Only at operational levels some concern is expressed and isolated initiatives are initiated. A more comprehensive and centralised structure is urgently needed to tackle this convergence issue. Furthermore, one could argue that different logics are guiding the EU policies in different hardware and software sectors. Therefore, the telecommunications policy with an emphasis on liberalisation and deregulation differs from the policy recommendations in the broadcasting field where some measures (e.g. the quota-system) could be interpreted as protectionist.

Other problems troubling the European Union concern the different policy consequences of economic vs cultural, national vs local interests. Especially in the area of broadcasting one should also point at the differences between hardware vs software, production vs distribution options. Some of these problems are generated from within the EU (as a result of fragmentation and bureaucratisation, for instance); others are conditioned externally (by US or Japanese policies). In general one could say that both national governments and the European Union as a governing body are faced with a dilemma when it comes to developing a communications policy. If they would give preference to economic and technical considerations, they would stimulate the media policies in the direction of uniformisation and large-scale developments. Quantitative criteria, which are mainly based on 'technical' (or hardware) considerations, do play a more important role than qualitative criteria that build upon the 'content' (or software) of media products. The latter approach would be more in line with a cultural policy, which emphasises pluriformity and small-scale autonomy (see also Becker, 1995).

Therefore, Van Cuilenburg and McQuail advocate a new Communications Policy Paradigm, which addresses new dilemmas and unanswered questions like: what balance to maintain between public and private sector; how to finance the public sector; whether and how far to direct policy and control content or structure? Such a new perspective, they argue, needs to focus on the media market structure rather than on media content. Burgelman takes a different position. He argues that communication policy has to shift from media policy to social policy. This implies that since communication policy has up till now traditionally been sector specific and mainly devoted to access to media matters or to the telephone network, regulating access in a more networked or information society should be pushed into new directions, both at policy and research levels.

In Servaes (1987, 1999) we have been trying to grasp the complexity of the above issue in a structural-dynamic way by integrating at least six basic elements into a policy and research model:

a. <u>content</u> related variables which can be of an <u>external</u> or <u>internal</u> nature (objectives, goals, resources and functions of the communications policy, etc.)

b. interrelated factors (economic, political, technological, socio-cultural, ... sectors)
c. actors (business, industry, governments, political parties, institutions, public pressure groups, influential personalities, etc.)
d. a distinction at the level of analysis (transnational, international, national, local)
e. the process of policy and planningmaking (policy formulation, planning, implementation, evaluation, etc.), and
f. the specific hardware vs software, and production-distribution-consumption problems inherent to any communications system.

From a '*structural perspective*' one can distinguish between actors and factors that influence policy and planningmaking. Actors can be defined as public or interest groups which, by both direct and/or indirect means, try to push through their explicitly or implicitly agreed policy objectives. Factors, on the other hand, determine the contours within which the actors can operate. Factors are time and place bound and differ from society to society. Some are of a conjunctural nature, others of a more structural nature. The drafting of policy objectives must be tuned to the needs and expectations of the various public groups and it must also take account of the resources available and the objectives of the communications system.

From a '*dynamic perspective*' one can distinguish between five overlapping phases in the policy process: the formulation of policy, the strategic or long-term planning, the operational or short-term planning, the implementation of the policy and planning, and the evaluation of it. Each level and phase of analysis of the policy process is determined by the dialectical intercourse of structural and dynamic components. Policy makers and planners have to solve problems at all these distinct levels.

However, as argued by Burgelman or Josiane Jouët and Sylvie Coudray (1990), much more research is needed in order to arrive at valid and realistic policy recommendations. The current state of knowledge on information availability and use by different sectors of society is extremely weak and fragmented. Furthermore, to be effective, the research community will have to extend the scope of its research and analysis to include implications of its research findings. It will have to pay more attention to the dissemination of its research results, and should at the same time become more involved in the policy development process.

Robin Mansell, David C. Neice
& W. Edward Steinmueller

Universal Access Policies for Knowledge-Intensive Societies

Introduction

Conventional views about universal telephone service are an inadequate foundation for resolving social and economic policy issues as economies come to depend increasingly upon knowledge-intensive activities. In this article, this position is argued first by considering the positions of proponents of the current approach to universal service, mainly in the European context. It is not argued that current universal service measures should be abandoned. Important issues continue to need to be addressed. There is a growing need, however, to clarify the nature of other aspects of universal service. These aspects involve social transitions that accompany innovations in communication technologies and services. The second half of the article deals with these social transitions.

Relevant social transitions in the universal service context include, first, the costs of skills and knowledge acquisition and, second, the implications of changes in perception of social status. This article argues that universal service in the context of network and service availability and affordability is an important element of policy especially in markets undergoing a transition from monopoly to competitive supply. However, as the new Internet-based mode of communication begins to complement the older 'telephony' mode of communication, new elements of the universal service concept need to be addressed and reflected in public policy. The two elements of a universal Internet service model, that is, skills and changes in social status perceptions, call for creative approaches to policy. This argument is consistent with the recommendations of the European Commission's High Level Expert Group report entitled 'Building the European Information Society for Us All' where it is argued that 'there is an urgent need to re-prioritise Europe's investment in education, training and human resources, and knowledge and skills acquisition more generally' (European Commission, 1997a: 23). This emphasis on skills and knowledge suggests that both public and private resources will be needed to finance universal access in settings where basic skills and knowledge needed for the new mode of communication can be acquired.

1. Accommodating technological transitions

Our modes of communication through geographic space and time have undergone a series of transitions throughout history (Innis, 1950). One major transition is represented by the succession of services available to the public which support personal and business communication. In a little more than one hundred years, we have moved from concerns about whether the telephone would substitute for postal services to debates about whether the Internet will provide the basis for a major shift of personal and commercial activity into 'cyberspace'. For the most part, complementarity, has characterised historical developments. Today, for example, while most postal services are diversifying into the supply of new electronic services, the physical delivery of letters and packages remains important to most aspects of commerce and trade. Although some services like the telegraph have been superseded by new text-based services, the historical trend has been to diversify the number of modes of communication rather than to move in a linear manner from one innovative mode of communication to another.

Each new mode of communication has been accompanied by its own set of policy problems. The problem is always to balance society's interest in benefiting from successive modes of communication with the limits either of public sector borrowing requirements or the private sector's interest in profits. Each of the predominant modes of communication has undergone a period during which service 'to the public' was limited (Davies, 1994; Marvin, 1988). Postal service was limited initially to the literate and wealthier classes of society, and telephony at the outset was regarded as useful only to a sub-segment of the social or business community. The growth of Internet access is posing new questions about how quickly it should become ubiquitous if the public's interest in the new mode of communication is to be met.

The ubiquity of service provision under the rubric of 'public service', and more recently, 'universal service' has varied historically as a result of changes in government policies, political priorities, and the economics of supply. There is no systematic relationship between universal service and the organisation of telecommunication service. For example, competition promoted the diffusion of telephony in the early days both in Europe and in the United States (Davies, 1994; Gabel, 1969). Mueller's (1993, 1997a,b) recent reconstruction of the history of telephony in the United States has confirmed this. The internal subsidies between the local, long-distance, business and residential, and between urban and rural services offered by telecommunication operators under both competition and monopoly have borne very little direct relationship to the achievement of universal telephony (Gabel, 1967). As Mueller argues, 'if competition actually *helped* to achieve universal telephone service and the Communications Act and cross-subsidies to local access had almost nothing to do with it, then the whole approach [today]... is wrong...' (Mueller, 1997b: 1-2).

As the penetration of telephony service increased, social and economic processes became more heavily dependent upon the use of telephony. The telephone service become deeply embedded in the distinctive cultures, social fabrics, and economies of nation states (Marvin, 1988). In the European countries and in North America, issues of universal service are bound up with the rights and expectations of consumers and citizens (Mansell, 1997). The absence of access, or even restricted access, to telephony services is associated with many different forms of social exclusion. For example, as Haddon argues:

> '...the phone facilitates participation in 'communities of interest' no longer based locally in the neighbourhood but where people living in wide areas come together regularly for leisure purpose or other interests...many have already started to perceive that their way of life involves some dependency on telephony for a variety of reasons' (Haddon, 1998a: 6).

On the other hand, however, people continue to seek ways of controlling and managing how and where they may be contacted as mobile telephony and distance working arrangements bring the telephone into the centre of their private spaces. Research by Haddon (1998b) and Ling (1997) shows that the rules governing arrangements for using the telephone continue to be ambiguous, subject to negotiation, and a source of tension and resistance even today. In the absence of systematic empirical research, the 'ubiquity' of the Internet as a new mode of communication cannot simply be perceived as a 'good' thing which should therefore be incorporated into the telephony model of universal service. The next section considers the current status of the universal telephony service model mainly in Europe.

2. The Universal Telephony Service Model

The universal service discussion of the 1990s basically revolves around questions of whether today's telecommunication operators can, or should be expected to, preserve or enhance their public service role (Rapp, 1996). Public and privately owned organisations are offering differentiated services that mix voice, data, audio-visual, and print-based modes of communication. The larger operators are seeking new global markets, alliances, and competitive service portfolios in liberalised, and largely privatised, markets. Convergence of the underlying technologies is transforming local and global markets by facilitating the variety of modes of communication which are available to consumers and citizens. At present a high premium is placed on achieving and maintaining universal voice telephony services.

The universal service obligations expected of dominant telecommunication operators have been addressed by the European Commission. A minimum set of services that should be offered to the public at affordable prices has been defined (European Commission, 1996b). One of the Commission's communications on universal service begins with the observation that 'to be able to communicate and interact whether by telephone, fax, e-mail or electronic media is a crucial and decisive factor for every citizen and business' (European Commission, 1996b: 1). This point of view has been accepted at the highest political levels (European Commission, 1997b).

The scope of universal service is understood as 'a defined minimum set of services of specified quality which is available to all users independent of their geographical location and, in the light of specific national conditions, at an affordable price' (European Commission, 1996b), and includes:

> '..the provision of voice telephony service via a fixed connection which will also allow a fax and a modem to operate as well as the provision of operator assistance, emergency and directory enquiry services and the provision of public pay phones' (European Commission, 1996b).[1]

This definition ensures that users will have a possibility of accessing services, including the Internet, that can be provided over existing telecommunication networks if they have a computer and a subscription to an Internet service provider (European Parliament, 1997). The delivery of universal service is associated with the duties of designated licensed telecommunication operators. Thus:

> 'In the new concept the role of the monopoly network operator is replaced by the concept of public network operators acting in competition, with a number of rights and duties. The rights concern the right to interconnection with networks of competitors of the same category. The duties concern the obligation to offer network access to others but also the guarantee of universal service' (Ungerer 1997: 2).

The Commission recognises that universal service must be a dynamic and evolving concept and that the liberalisation of member state telecommunication markets will impact on how this concept evolves. A study by Analysys (1997) establishes the penetration rate for any new candidate for 'universal' service at 75 per cent of the potential user population. At this threshold, only mobile services might be considered to fall within the Universal Service Obligation by the year 2005. The Commission also has determined that 'voice telephony' service offered by Internet service providers is not a candidate for consideration under Universal Service Obligations that are currently in place (European Commission, 1998a).

The dilemma for policymakers is that moves to stimulate investment in more sophisticated facilities in advance of evidence of market demand could result either in a stimulus to demand benefiting society as a whole or the construction of wasted capacity and content at costs that might jeopardise the economic interests of suppliers (Analysys, 1995; Blackman, 1995). In line with the Commission's goal of stimulating competition in the European telecommunication market, universal service has been considered in terms of the costs that incumbent operators and new entrants claim are associated with its provision. A methodology has been decided for costing universal service obligations. Funding mechanisms authorising the sharing of identifiable cost burdens associated with the provision of universal service have been established (e.g. a national universal service fund or a system of supplementary charges requires from market players by the incumbent operators) (European Commission, 1996c,d; 1997c). In practice, however, 'a majority of Member States are either judging any burden associated with universal service to be so low as to not constitute an unfair burden for the operator concerned or are proposing the creation of national universal service funds' (European Commission, 1998b).

The focus of the universal telephony service debates has been on access to networks, services and affordability. Thus:

> 'The Directives foresee that a set of basic services should be 'affordable'. There is a European list of such services, but the notion of affordability itself varies from country to country since purchasing standards are different. In those cases where the provision of these services brings a net cost to the operator providing it, contributions can be raised from other operators: this is called the financing of universal service ... Telecom's liberalisation is not only a change of regulations. It is also a change of mentality, both for operators and for consumers' (Van Miert, 1997).

The Commission's 'Green Paper on the convergence of the telecommunication, media and information technology sectors, and the implications for regulation — towards and information society approach' provides a basis for further discussion on the evolution of universal service concepts. It acknowledges that network convergence (the ability of different network platforms to carry essentially similar kinds of services) is becoming more commonplace. It foresees the integration of consumer devices such as the telephone, television and the personal computer from the point of view of consumers (European Commission, 1997d). It suggests that the public's interest in 'general interest services' and 'universal service' will continue unabated. It also suggests that problems of defining obligations and the appropriate cost basis for determining the burden they place on operators are likely to worsen.[2]

As we move away from the universality discussion in the context of telephony towards a discussion on consumer expectations and rights in the information society, there is considerable support at political levels in Europe for the idea that people need access to the Internet, but there is a bifurcation in the views about how this should be achieved. On the one hand, for example, the European Commission's Working Group on the Cultural Dimension and the Future of the Media of the Information Society Forum has argued that:

> 'Even if the economic outlook for new services appears to be doubtful at present, members of the working group have spoken out against the public funding of commercially operated services. The market should determine the conditions for the success of these services' (European Commission, 1996e: 95).

On the other hand, the Forum's Working Group on Basic Social and Democratic Values sees scope for policy intervention to stimulate market development:

> 'Citizens must have the possibility of enhanced participation in democratic structures as well as access to a wide variety of public information, entertainment, retail, education and health services. These should be provided for all. The Working group believes that the development of networks and operating systems must enhance access and choice and must not exclude certain groups in society' (European Commission, 1996e: 21).

In the United States, the main points of the universal telephony service debate are not significantly different. The Telecommunications Act of 1996 produced renewed debate on the scope and delivery of universal communication service. Although the concept was extended to services for schools, healthcare facilities and libraries for the first time, the definition is driven by politics, technical innovation and trends in market demand just as it is in Europe. The impact of the new legislation in the United States will depend on how it is implemented. The definition of what is included in the universal service category is substantially the same as in Europe.[3] The Joint (Federal / State) Board on Universal Service established by the Telecommunication Act of 1996 leaves judgements about the affordability of services to the State jurisdiction (Federal Communications Commission, 1996), just as this is left to the member states in Europe. The Board is empowered to recommend funding support for low income consumers. The Act provides a basis for determining the amount of support that should be available to carriers with high costs, but it has not mandated that network and service operators should pay costs in excess of the underlying costs of supply. This approach is similar to the cost-based approach that has been adopted in Europe. Federal funding programmes in the United States are aimed at promoting private sector investment in new infrastructure and services, not at fundamentally altering the consensus on what consumers and citizens 'need' in terms of basic service availability and affordability.

The focus in both Europe and the United States on access and the affordability of networks and services is insufficient however for the governance of the new mode of communication that seems likely to complement telephony in the 21st century. Although both Europe and the United States are addressing the universal service issue in similar ways, the policies are a reflection of the 'telephony' model which has served relatively well until now. During the transition to a new mode of communication, however, a new model is likely to be needed. The following section considers two necessary elements of a new 'Internet' universal service model.

3. Towards new models of universality — The Internet Mode of Communication

The Internet allows users to be information collectors and providers simultaneously (Rheingold, 1993; Steinmueller, 1992; Guena and Steinmueller, 1997) and this is a distinctive aspect of the new mode of communication. Any user can set up a Home Page on the World Wide Web and many Internet Service Providers (ISPs) are information providers. The Internet's ability to support both the generation and supply of information in a multimedia format means that its economic value to users is likely to increase more rapidly than a telephone network as the network dimensions increase. However, Internet use depends fundamentally on the user's capability. Internet functionality is diverse, including interactive communication, information dissemination, and information collection. Intelligent agents that search for information selectively have no equivalent within telephone networks and the extent of information resources available through the Internet exceeds any other medium for communication.

Over 25 years, the Internet has developed from a loose federation of networks based on non-user-friendly software to a global communication network. The Internet has doubled in size every year since 1988 and currently has around 65 million users. A wide range of suppliers, including Internet Service Providers and software developers, are interested in shaping the Internet's future development (Narendran, 1997). How can provision of Internet services be secured for a wider range of users and under what conditions should they be supplied? The importance of the accumulation of skills and capabilities appropriate to the new mode of communication is being recognised, but policy debate continues to focus on the economics of network and service provision. For example, in the United Kingdom, the out-going Director General of Telecommunications, Don Cruickshank has commented that:

> 'It is important that schools should be able to get access to the same high quality content for the same low cost, no matter where they are. It would be wrong for a school in a remote area to have to pay more, just because connecting them is more difficult. We want a society in which all children get equal opportunity to make the most of the Information Age' (Office of Telecommunications, 1998).

The new mode of communication gives rise to the need to increase access to opportunities for skills accumulation rather than simply to affordable network connections and services. This raises new and urgent issues for universal 'Internet' service policy. The European Commission, for example, is emphasising the need to equip schools with multimedia and audio-visual tools and to develop teaching resources and provide training and logistic support for teachers (European Commission, 1996a). The provision of 'universal' services for education and training relevant to building Internet capabilities will be as important in the 21st century as the economics of network access provision. As the report of the High-Level Expert Group of the Commission's Employment and Social Affairs Directorate points out:

> 'If we accept the argument that developing technological capabilities does involve a complex, endogenous process of change, negotiated and mediated both within organizations and at the level of society at large, it is obvious that policies cannot and should not be limited to addressing the economic integration of technological change, but must include all aspects of its broader social integration' (European Commission, 1997a: 18).

3.1 Models of Internet Skills Acquisition and the Accumulation of Capabilities

The cost of social exclusion needs to be considered in parallel with the commercialisation of the Internet. Even if telephone network operators and Internet Access Providers offer unlimited access to the Internet over a standard telephone line for a flat fee, this will not fully address concerns about social exclusion. This is because the transition to a new mode of communication, i.e. the Internet, implies a transition to a new skills base as well. Users need to be computer network literate (Lanham, 1995; Gilster, 1997), comfortable with using computer interfaces, and they need access to greater levels of support than is the case for the use of voice telephony services. Skills and the accumulation of new Internet capabilities are associated with consumer and citizen empowerment about choices concerning their selection of services in the information society (Graham et al., 1996). The following discussion illustrates the close link between opportunities for skills and knowledge development and the types of organisational settings where it is possible to acquire those skills (Guena and Steinmueller, 1997).

The typical Internet connection used by large organisations is based on a leased line, usually a T1 (E1) line, and a local area network that links terminals and personal computers distributed throughout the organisation. The management of the system, the supply of services, and the technical and training support are run by a specialised department or team within the organisation which may, in turn, outsource parts of its responsibility to outside contractors. In this case, as Guena and Steinmueller (1997) illustrate, Internet access costs are composed of the subscription to the service provider, the rent of the line, a fraction of infrastructure costs for PCs and the local network, and a share of the management and training costs.

The final user, the employee, is unlikely to bear any of these costs. In some cases, employees will invest independently in their skills to improve their performance. In the organisation, the costs of providing Internet access (including skills and knowledge development) are substitutes for other investments. The use of personnel who are knowledgeable to provide Internet access is a substitute for using these individuals to perform other functions within the company. The growth of 'Intranets', private networks using tools and techniques similar to those used on the public Internet, is very rapid. These opportunities are leading to a growing familiarity with the tools and methods of Internet use and a potential spill-over to the use of the public Internet. Larger organisations should be expected to continue to make significant contributions to the development of Internet-related skills and knowledge over time. They have a significant role to play in future universal 'Internet' service policies. As Guena and Steinmueller (1997) comment, although this contribution to a common pool of skills for using the Internet is valuable, it raises questions about growing imbalances in the acquisition of these skills.

Both home and small enterprises use the Internet via dial-up access. Telephone and Integrated Services Digital Network connections supplied by public network operators connect users to an Internet or on-line service provider offering access to Internet and to other services. Internet access costs are composed of the subscription to the service provider, the usage costs for the network connection, a fraction of the hardware costs (usually a PC), and skills acquisition and support costs. An extremely important component of this cost structure is the personal investment the user has to make to learn how to use the system. Compared with the use of the telephone or television, using the Internet is complicated and substantial knowledge and skill are necessary at the beginning. These costs differ in relation to the age and education of the user and they depend on the presence of a support structure. The existence of an education system that familiarises students with the new technologies can reduce the costs that must be paid to achieve access (Guena and Steinmueller, 1997). Improvement in support services, from service providers and hardware sellers, also represents a market opportunity that could be supported by public policy.

The cost structure of community access to Internet services is similar to that of individual access, but the components have different weights. Only the largest public schools and libraries generally have the financial resources to consider a high-capacity leased line as a means of acquiring service. The school, library or museum of average size is likely to access on-line services through the public switched network and to provide access for others through 'host' service providers. Since schools and libraries have the highest use of communication during the peak daytime hours, the usage costs of connection are an extremely important element in the cost structure. There are also significant costs of employing qualified people who can provide training and service support.

In each of these Internet models — organisational, individual and community — there are two elements in addition to the availability and affordability of telecommunication access. One is the cost of equipment for the user and the other is the cost of developing skills and knowledge. The first may be improved by the development of equipment dedicated to Internet use. The second is likely to continue to present a major barrier to the ubiquity of Internet usage. This is an area that could be addressed systematically by public policy. Funding this component as a part of basic education would reduce private training costs and also create substantial spill-overs as employers find it easier to attract individuals with basic computer skills.

The most important next steps for the Internet universality agenda involve not only the gradual redefinition of 'universal service' in terms of the economics of network supply and use, i.e. within the context of the telephony universal service model. They involve new universal Internet service policies to encourage the acquisition of skills and experimentation. This will require specific cost estimates for attaining access to new skills-building opportunities, debate on the merits of undertaking this goal from a social viewpoint, and better assessments of the social implications of Internet as a predominant mode of social and business communication.

3.2 The Internet and the Social Order

In addition to the costs associated with skills acquisition, there is a growing need to link the implications of the new mode of communication to transitions in the social order and the potential for social exclusion. Empirical research on the social implications of the Internet as a transitional mode of communication is crucial if effective universal service policies for the 21st century are to be developed. What exactly is the public's interest in broadening the base of user skills and knowledge for Internet participation? At present answers to this question rely mainly upon anecdotal evidence and studies commissioned by those seeking to market Internet-based products and services.

Freeman and Soete argue that the public's interest is substantial in this area. In their view, social exclusion is a serious threat when people cannot achieve effective access to the new modes of communication.

> 'A fairly large number of people, even in the richest countries, are unable or unwilling to use the new technologies or to gain access to facilities where they might be used. 'Information poverty' corresponds fairly closely to material (income) poverty but is not identical. Social exclusion and the growth of a large underclass are thus becoming characteristic features of the information society ...' (Freeman and Soete, 1997: 410).

Others have pointed to the economic and social divides that innovations in computer networking and the commercialisation of services based on the Internet may be generating (see Handy, 1990; Drucker, 1993; Romer, 1993; and Lipsey, 1995). Until the 1990s, Neice (1998) argues that individuals had a reasonably clear understanding of the main facets of social privilege and of social inequality. Occupation, gender, ethnicity, ownership of specific types of property and styles of consumption indicated where most people were positioned on a 'map' of social status. He argues that changes in the way that social status is signified may be accompanying the diffusion of Internet-based services and applications. The universality of Internet use takes on major importance in this light. For many observers, the Internet appears to be a levelling device that is expected to give rise to a more egalitarian social order. However, this perception is inconsistent with current Internet user profiles and with studies of the many ways in which social exclusion can occur:

> 'While societies as a whole become progressively more dependent on electronic media and information and communication services, then those who find access, for financial or other reasons, increasingly difficult or just simply impossible will become progressively marginalised and excluded. Isolation can take many forms' (Mansell and Silverstone, 1996: 224).

The social science literature on the associations between the widespread diffusion of a new mode of communication and tendencies for new patterns of social and economic exclusion to emerge offers insights into the social and economic transitions accompanying technical innovations in advanced information and communication technologies. One transition involves a growing trend toward the dematerialisation of many social and economic activities. 'Dematerialisation' refers to the generation of economic value based upon an increased valorisation of 'bits' and knowledge in contrast to 'mass' and material things (Romer, 1993; Negroponte, 1995). Quah (1996) suggests that a dematerialised economy is one in which bits and bytes congealed in software or gene sequence codes increasingly become the primary components of world trade. In the emerging economic order, he suggests that 'subgroups in societies form knowledge-based coalitions ... Over time, small differences across social groups magnify ... patterns of polarisation and stratification appear' (Quah, 1996: 9).

Similarly, Castells (1989, 1996) sees the emergent 'informational mode of development' as leading to new economic growth opportunities and the internationalisation of the economy. This second transition involves new forms of knowledge-intensive activity. The informational mode is characterised by increased concentration of knowledge generation and decision-making processes in high level organisations (Castells, 1989: 348). As physical places become subordinate to the logic of 'cyberspace' or the 'space of flows', opportunities for new social and economic divisions emerge. Building skills and capabilities to engage in social and economic transactions in virtual communities is essential to creating new pathways towards social inclusion (Castells, 1997).

The social stratification of society has been defined as 'a hierarchy of distinct social groups, each having specific life chances and a distinctive style of life'(Scott, 1996:1). The hierarchical divisions that mark identifiable status groups are subject to cultural variability and change. Analysis of the potential changes in social status 'maps' accompanying the rise of a new communication mode requires detailed examination of transitions in the perception of social status, especially among those who are participants and non-participants in Internet's new virtual communities (Steinmueller, 1998). Empirical research is needed on changes in social status perceptions within and outside the new virtual communities (Neice, 1998). The Internet is a useful exemplar of these issues because it requires skills in computer-mediated communication. These skills may change as technologies and services evolve, but they provide an initial basis for determining whether their presence or absence is associated with new divisions within (Neice, 1997).

Evidence from marketing and opinion survey studies in North America suggests that demographic groups and social strata gain experience of the Internet very unevenly (Neice, 1996). North America's Internet users are mostly young and male. They are relatively affluent, they have extensive education, live in large cities, are consumed by the work ethic and they are strong believers in the benefits of applied technology. High levels of participation with digital networking tools involving ownership and use of personal computers, modems and associated digital peripherals (CD players, cell phones, etc.) are present among these users in both North America and Europe (European Commission, 1997b).

Evaluating social status and position in societies characterised by greater knowledge intensity is very difficult because we have yet to develop appropriate measurement tools and indicators. The new forms of social stratification and exclusion appear on the basis of preliminary evidence to be linked not only to the standard list of property ownership, income and life-style choices, and occupational position or location, but also to skills and capabilities required for membership in the new stratum of knowledge workers. Access to information and knowledge resources and to the Internet, ownership or lease of equipment, and the ability to produce new digital information are symbols of membership in the newly emergent social order. Social positioning appears to be related to a growing extent to three types of access: material and physical access to infrastructure networks, hardware, and support resources; access to certain knowledge- or design-intensive occupations and the requisite specialised technical training, skills and capabilities; and access to social groups and communities which reinforce formal skills and facilitate the accumulation of the knowledge base (Neice, 1998).

If access to the new mode of communication is entwined with perceptions of hierarchy and status in the evolving social order, then universal service models will need to incorporate an understanding of the skill requirements for digital literacy as well as the symbols which characterise new patterns of social stratification.

Conclusion

The traditional approach to universal service in the context of competition policy continues to be important. This is especially so as differences between the major trading regions of the world persist in the technical availability of networks and the costs of their use. At present, for example, in most member states of the European Union the penetration of telecommunication and multimedia services has not achieved the levels of the United States (European Commission, 1997e). An OECD study shows that 'the penetration of Internet hosts is five times greater in competitive than monopoly markets, and if allowance is made for the date of service commencement, Internet access in countries with telecommunication infrastructure competition has grown six times faster than monopoly markets' (OECD, 1996a). This suggests that competition may be as effective in stimulating the new Internet-based mode of communication as it was in the early days of telephony. Nevertheless, policies to monitor the 'universality' of basic connectivity will continue to be important in the context of the creation of a level playing field for competition and to ensure that basic facilities are affordable.

This article has illustrated the importance of two additional aspects of the new mode of communication which are essential aspects of the new governance models that will be needed in the 21st century. The first is the importance of learning and skills acquisition. Universal service policies will be needed to ensure that people can acquire skills and become familiar with the new mode of communication. The second is the transition in the underlying patterns of social division associated with the emerging symbols of participation in knowledge-intensive economies. Existing empirical evidence so far provides only hints of emergent trends. It will be essential, however, for new governance initiatives to reflect the transitions in the social order. Governance in line with the new elements of universal service is unlikely to fall within the purview of the regulatory institutions with remits for implementing the universal telephony model, but new policy initiatives for universal 'Internet' service models will need to be closely linked to the existing model. Promoting access to the Internet, in the absence of new policy measures with respect to both skills and the divisions created by changes in the perception of social status, could simply engender more deeply rooted patterns of social exclusion.

The producers of electronic commerce and 'general interest services' in the private sector have a clear economic interest in contributing to new universal Internet service policies because this will strengthen the pool of skills and capabilities of users and encourage market development. Governments have a strong interest because of their concerns about citizen participation in a civil society which relies increasingly upon a new mode of communication and their interest in reducing the negative impacts of social exclusion. A new rationale for 21st century universal service and a new governance system with a broader remit than that traditionally associated with universal service are therefore likely to have broad appeal.

Notes

1 Voice telephony service is defined as the commercial provision for the public of the direct transport and switching of speech in real time between public-switched network termination points, enabling any user to use equipment connected to such a network termination in order to communicate with another termination point.

2 For example, 'Video-on-Demand is either treated as a value-added telecommunications service or has not been formally categorised in all member States, except France and the United Kingdom where it falls under the broadcasting framework and Germany where it is under a new category 'teleservices', (European Commission 1997d: 54 (fn 48).

3 In the United States the definition includes voice grade access to the public-switched network, including, at a minimum, some usage of dual-tone multi-frequency (DTMF) signalling or its equivalent; single party service; access to emergency services; access to operator services; access to inter-exchange (long distance) services; and access to directory assistance.

Jan van Cuilenburg and Denis McQuail

Media Policy Paradigm Shifts: In Search of a New Communications Policy Paradigm

1. Three phases of media policy

The origins of communications policies lie in the interaction between the pursuit of national interests by states and the operations of commercial/industrial enterprises. Both government and industry have sought mutual advantage by way of privileges, regulations and restrictions. Policies in general refer to conscious (public) projects for achieving some goal, together with the proposed means and time schedule of achieving them. The specific content of government policies reflects the deal made in the particular time and place and the balance of power and advantage between government and industry. At this level of generality, not a great deal has changed, despite the general transition during the 20th century to more democratic forms of government. Even so, in the field of media policy there have been significant developments, especially reflecting the changing technologies of communication and the increased importance of communication in 'post-industrial' societies.

Although for centuries there have been state interventions and regulations relating to transport and travel for purposes of control, finance or for strategic reasons, it would be anachronistic to speak of communications policies before the series of electronic inventions beginning with the electric telegraph in the mid-19th century. In this article, we identify three main phases of communications policy-making. We may label these consecutive phases as (I) the *Phase of emerging media policy*, (II) the *Phase of public-service media policy*, and (III) the *Phase of communications policy.*

Phase I: Emerging media policy (... - World War II)

The first phase is transitional from the time of 'no-policy' and consists of a collection of ad hoc measures designed to regulate and facilitate the introduction of a series of innovations, from the mid-19th century until approximately the start of World War II. This phase is characterised by piece-meal accumulation of measures, with varying aims, means and scope. There is no coherent goal, beyond that of protecting or advancing the interests of government and nation, and promoting the development of communication systems, whether by state or private capital investment. During Phase I, the lines were laid down for later policy and law for telegraph, wireless, film, television, cable transmission, phonography and various related media. A principal characteristic, much remarked on since, was the separation of regimes for different technologies, with particular reference to the means of distribution. This separation was due more to historical accident than to design.

Phase II: Public-service media policy (1945 – 1980 / 1990)

The second phase of media policy extends from after World War II until about 1980. It was characterised more by normative and political than technological considerations and by the search for national coherence and stability. Despite the influence of the Cold War during this period, the spirit of democracy and the wish for international solidarity were strongly expressed in policy. The lessons of the misuse of mass media for propaganda were not completely forgotten. There is no precise point of conclusion, but the key factor making for a new phase was the influence of computer technology on telecommunications and the practical application of satellite communication for private purposes, including mass communication. The stage was set in the early 1980s for rapid expansion, transnationalisation and *convergence* of technologies.

Phase III: Communications policy (1980 / 1990 - ...)

Convergence became an agenda item when the former US *Office of Technology Assessment OTA* published its pioneering study *Critical Connections* (1990). Later in the decade, the European Union adopted the same theme for its new approach to communication policy (European Commission, 1997d). Technological convergence means, for instance, that the boundaries between information technologies and communication networks are technologically blurring: computer and telecommunications are converging to telematics, *PC* and television sets get more and more resemblance, and formerly separated networks become more and more interconnected to render the same kind of services. Multimedia, integrating text, audio and video, also is an example of technological convergence. Next to technological convergence, we currently see economic convergence in the communication and media sector, that is the merging of the branches of computing, communications and content (publishing). It is because of technological and economic convergence that the idea of 'communications policy' was born and took a clear shape. Ministries of communication were founded and new media laws promulgated. Regulation on media became increasingly connected to telecommunications regulation. In addition to technological changes, social-cultural changes are central to our time too. The 'decline in ideology' which has been reflected in the fall of communism and the increased scope and respectability of the free market is a notable feature of recent developments. Globalisation of communication and the permeability of national frontiers by multinational media are also new in the scale of their impact.

The general shape of communications policy

The main elements or factors of media policy and telecommunications policy consist of the *goals* to be pursued, the various *content and communication services* (information, culture / entertainment and common carriage services) that

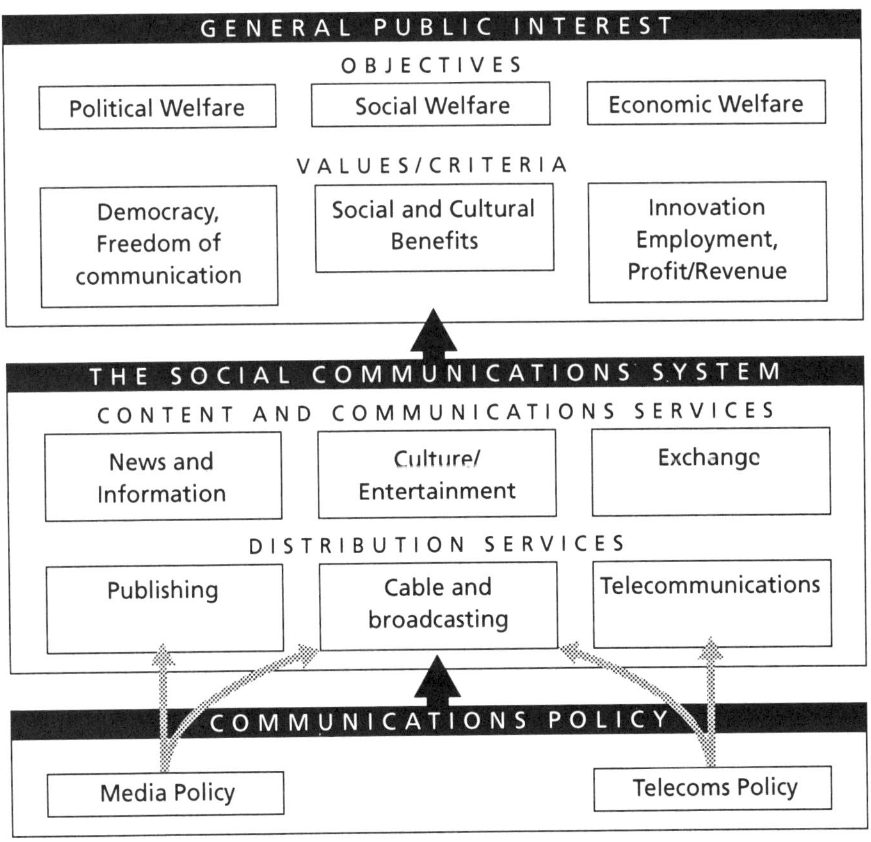

Figure 1

are provided, and the *distribution services* (publishing, cable and broadcasting, and telecommunications) involved. The relationship between these elements in a given national environment is sketched in, adapted from Van Cuilenburg and Slaa (1993, 1994).

Figure 1 depicts that communications policy includes both media policy and telecom policy. Traditionally, in most countries media policy focuses on the 'left' part of the model (publishing, CATV and broadcasting), whereas telecommunications policy by definition is directed at the 'right' part, mainly telecommunications and to a lesser degree cable and broadcasting. In the model, the social communications system itself is represented as to be tri-functional, that is, is considered to serve political, social and economic objectives in the general public interest.

The model is a view of policy as formulated by *government*, rather than by media corporations, although we have to recognise in the present phase, that the business strategies of these corporations may be more influential than the policies of some governments. As far as this influence has to be represented in a model of policy of this kind, it mainly concerns three things. These are the interpretation given to national economic welfare; the balance between services provided for private profit or other social purposes; the ownership of distribution systems and infrastructure.

2. Phase I: Emerging media and telecommunications policy (... - World War II)

In the early phase of emerging communications policies especially as instanced in the United States, Europe and the British Dominions, the political functions of communications were largely unrecognised and unregulated. The common assumption was that the medium for overtly political communication should be the printed newspaper, for which there was no policy except freedom from censorship and subjection to the rule of law of the country in which they operated. In Phase I, policy refers primarily to the emerging technologies of telegraph, telephony and wireless, although the cinema was also soon regulated when it arrived at the turn of the century.

2.1 US telecommunications

There is clear evidence of policy in relation to the electric telegraph, telephony and wireless. In the United States, the telegraph rapidly developed into a private monopoly, largely owned by Western Union. Telephony, when it arrived, was also virtually monopolised, apart from local operations, by another large corporation, AT&T. When Western Union and AT&T merged in 1913, an agreement was reached with government, accepting some separation of services and regulation in return for effective monopoly over long distance telephone and telegraph (Sussman, 1997: 77). Aside from brief wartime intervals, this model of government-regulated private monopoly was the core of early communications policy in the United States. However, the emphasis was on private ownership, and regulation was prompted more by general anti-trust sentiment and support for competition rather than by a vision of service to the public interest. What was good for capitalism was good for America, that is, private industry operating in a competitive anti-trust environment was considered to be the most important mechanism for high quality and innovative telecommunication services. Nevertheless, the US postal service remained a government preserve, a reminder at least of alternative structures.

2.2 European approach

The main alternatives were in evidence in Europe. In Britain, law reserved all telegraph services for the Post Office as early as 1868 and a similar pattern was followed in Germany, France, the Netherlands and Sweden. Aside from minor experiments and deviations, much the same can be said of telephony. The European policy was to make telegraph and telephony, along with postal services, a public monopoly and effectively a branch of government. Wireless communication followed the same path in its turn in the early 20th century. Even in countries firmly dedicated to capitalism, state monopoly was regarded as the best way to serve the national interest (as defined by the state itself). The electronic media engineering and infrastructure were seen as too strategically essential to the state and to the industry to be left to the uncertainties of the free market. They were scarcely regarded as consumer goods and services. They were also generally regarded as either 'non-political' or outside the scope of democratic political debate, a perception supported by their status as branches of the civil service. Presumably, the imperial(ist) traditions of the main European powers helped to produce and protect this general policy model.

2.3 Common features

It is not easy to fit the early history of American and European policy-making within the same framework, since the means of policy, in terms of ownership and control seem so divergent. The American model also led to a more rapid expansion in popular use of the 'new media' of the age (the telephone in particular). Both the 'public service' model of the European PTTs and the American capitalist model promised a principle of universal access, although the latter delivered more of the reality. Nevertheless, both models shared certain broad features, namely:

- treating communications media as branches of industry with strategic importance
- separating out different sectors for medium-specific regulation (different regimes based on technology)
- subordinating communication to the imperative of national interest.

These features are logically related in that they each treat the field of communication as an essentially technical matter that should be settled by experts, lawyers and administrators rather than by democratic political choice. There is little reference to the *content* of communication in policy or regulation. They are effectively kept out of the sphere of public debate. While this reflects a general purpose on the part of business and government of keeping close control over matters of strategic importance, it also reflects a lack of social-cultural content in policy for the then new media. The policy terrain was effectively apolitical.

2.4 Cinema and Radio

A somewhat different assessment is required for later additions to the spectrum of early 'new media': cinema and radio. In both cases, a social and cultural dimension does develop, although it is not initially the driving force of government concern. The cinema was first treated in a similar way to the theatre and other public entertainment and shows. The main aim of public policy was to ensure safety (fire regulations governing public exhibitions and buildings) and then prevent potential harm to morals and the young by way of various forms of censorship and control. Regulation was typically on a local or city basis. Pressure groups of churches and concerned citizens led the way in this second goal of media policy, the prevention of harm to morals and the young.

It is worth noting a third goal of media policy in the Phase I era. Official governmental departments of propaganda were often behind the making of films and, later, news film (especially in war time). In some countries, notably Soviet Russia after the revolution and, to a somewhat lesser extent, in Nazi Germany, conscious efforts were made to control film production. In these countries the use of film and all other mass media was part of a general propaganda policy for promoting the values and ideology of the regime and suppressing criticism.

The case of radio provides a bridge to the second phase of policy. Although public radio (broadcasting) was perceived initially as an industrial and commercial device, with regulation designed initially to establish order in the market, set standards and safeguard official uses of airwaves, the goals of policy soon changed. During the 1930s, there was much legislation and regulation in North America and Europe which governed who should have access to the new channels and also for what purpose. Control of content was much stronger where public ownership was chosen above commercial exploitation of the new medium.

3. Divergence of communications policy 'regimes'

The separation of different media fields in terms of regulatory regime into three main distinct spheres, defined primarily by technology, dates from the later years of Phase I and follows much the same pattern in many countries. The three spheres (following Pool, 1983) are as follows:

1. that of print media, governed by guarantees of freedom of speech and expression (absence of censorship), subject to the law. This sector is characterised by voluntarism and private initiative, with the role of government limited to marginal issues. In the first phase of policy neither the structure of the press nor its quality was regarded as within the competence of government
2. that of common carriers, primarily telegraphy and telephony, with strong regulation of ownership and infrastructure, but not of content, and accompanied by certain rights of general public access and of privacy. The public mail service provided the basic model for this and efficiency, good administration and a certain concept of 'public service' were the hallmarks
3. that of broadcasting, first wireless and later television, along the same lines. Despite divergences between Europe and North America in patterns of ownership and forms of financing, there gradually emerged (in the later 1920s) certain common features. These included strong regulation of access and content, restricted freedom of expression, some form of monopoly or oligopoly, a notion of public service (in socio-political terms) and pressure towards universal provision. The further development of this mode of control takes us into the second policy phase that we have identified.

As Winseck (1998b) has argued, the 'divergence' of regimes, although seeming to follow demarcation lines based on technology with some apparent logic, is actually not itself a necessary consequence of technology. There is no intrinsic reason for the variations in the degree of freedom or the strict allocation of different means of distribution to different forms of control. The regulatory configuration that had emerged by the end of Phase I was not planned in advance, although there was some logic of precedent (e.g. using the mails as a model for telegraph and telephone). It was really the outcome of the pursuit of self-interest, sometimes diverging and converging, on the part of large industrial interests and the state. So, accident and historical circumstances rather than conscious policy seem to account for the separating out of policy regimes.

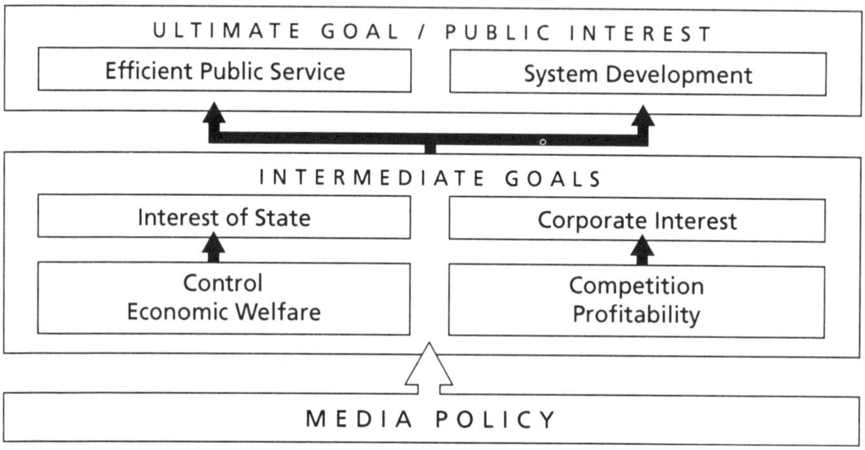

Figure 2: *Phase I* (... - World War II) The paradigm of emerging media policy

The main features of the rationale for policy in Phase I can be summarily presented as in. Media policy is pursued in the public interest that is restrictedly defined as efficient public service by and technological and economic development of the communications system. Against this background, the main criteria for media policy are the interest of the state and the financial benefit of corporations. These are measured in terms of control and economic welfare on the one hand, and by competition and profitability on the other.

4. Phase II: Public-Service media policy (1945–1980/1990)

The postwar (World War II) phase of media policy was dominated by sociopolitical rather than economic or national strategic concerns. The impetus to a much more self-conscious type of media policy (compared to Phase I) had several origins, but most generally it reflects a more collectivist spirit and the maturation of the politics of full democracy, which developed only after the turn of the 20th century in many industrial countries. The maturation process involved a more adequate evaluation of the significance of mass media for political and social life in 'mass democracy'. Early 20th century experience taught lessons about the power of propaganda not only in totalitarian society but also in capitalist societies dominated by large companies with 'press barons' as their allies. In some countries, notably Britain and the United States, the press by mid-century seemed less a healthy component of political life than an instrument unfairly biased against labour and progressive reform.

In respect of broadcasting, the spirit of social reform, encouraged by post-war 'reconstructionism' emphasised the positive social benefits of broadcasting run for public service goals, a well as offsetting the unfair political bias of capitalist newspapers. On the continent of Europe, the end of World War II provided the opportunity and sometimes the necessity for reconstructing the entire media system on more democratic lines after war, occupation or dictatorship. Even the once-sacred print media could legitimately be brought within the scope of policy. The general spirit of the time was favourably disposed to progressive change and to social planning in all spheres of life.

4.1 US Commission on Freedom of the Press (1947)

The ideas needed to give shape to policy were already available, especially in the form of earlier democratic theories of the press, including a critique of monopoly capitalist control. The early phase of public service broadcasting, which emphasised social and cultural objectives rather than entertainment or profit, helped considerably to extend the scope of policy. Further ideas were provided by the influence of the 1947 American Commission on Freedom of the Press (Hutchins, 1947). This made a clear link, for the first time, between freedom of the press and 'social responsibility', meaning an obligation to provide trustworthy and relevant news and information as well as opportunities for diverse voices to be heard in the public arena. The idea that freedom of expression carried some kind of social duty was controversial and resisted by newspaper owners, but it was nevertheless influential in affecting the longer-term climate of opinion. In the United States, this made it easier for the controlling body of broadcasting, the FCC, to impose requirements for fairness and diversity (e.g. as in the 1946 FCC 'Blue Book', cf. Krugman and Reid, 1980). The promotion of the virtues of 'freedom' in the context of the Cold War and the global battle for hearts and minds also made the notion of a democratic and public-spirited policy more compatible with an ideology of free enterprise.

4.2 Europe

In Europe, the traditional political bias of newspapers (whether to left or right) coupled with growing tendencies towards economic concentration in the postwar period provided the incentive for scrutinising the press (e.g. the Royal Commission on the Press, 1947-49 in Britain or the Swedish Press Commissions of the 1960s (Hulten, 1984)). In several countries, pressure was brought to bear (by various means, including legislation) to promote diversity of ownership and content, limit monopoly and deal more effectively with complaints against the press. An emphasis on *positive* rather than *negative* freedom of the press was an important element in what Picard (1985) has described as a 'democratic socialist theory of the press'. In general the various measures taken to maintain press diversity all embody the spirit of a 'free and responsible press' as a guideline for policy. The generally more corporatist traditions of European government allowed media policy to go much further than in the United States. This was particularly apparent in broadcasting. For three decades after the War, most European countries operated a public monopoly of radio and broadcasting, with Britain the only significant (and partial) exception. Decisions about the expansion of broadcasting and its broad tasks were political decisions and broadcasting (private as well as public) was made accountable for the nature and quality of the services it provided, despite often having considerable editorial independence.

4.3 Public-Service broadcasting

The obligations laid on public-service broadcasting do vary significantly from one system to another, but generally they involve the following elements:

- a commitment to universal service
- diversity and representativeness of content in political, social and cultural terms
- democratic accountability
- significant elements of public financing
- non-profit goals.

For much of the postwar period — the Phase II referred to above — media policy in Europe was largely co-extensive with policy for public broadcasting. The character of this policy began to change from the mid-1970s. It became increasingly pre-occupied with the financing and viability of public broadcasting, with preserving or removing its monopoly status and with the problems of incorporating new media developments into existing systems of regulation. This last applied in particular to cable transmission, trans-frontier satellite broadcasting and the new 'telematic' media of videotext and teletext.

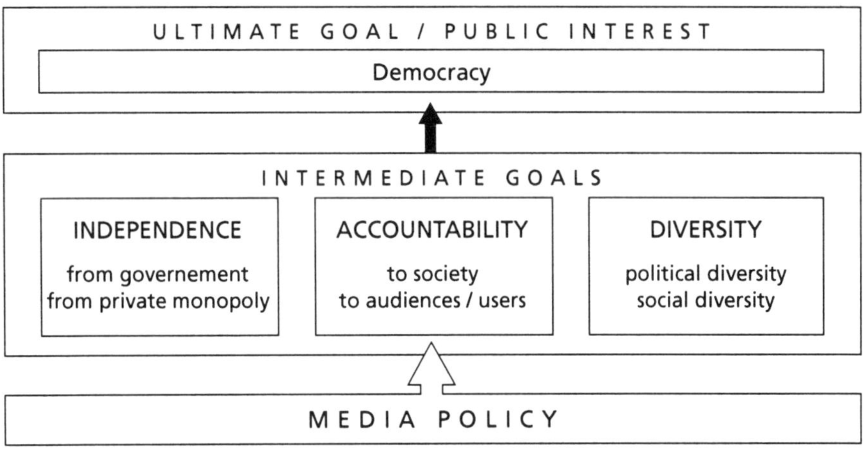

Figure 3: *Phase II* (1945 - 1980 / 1990):
The paradigm of public-service media policy

The chief contours of the rationale for media policy in Phase II are set out in Figure 3. We can summarise the main features of the public-service media policy paradigm in terms of the following main points:

- it is primarily shaped by normative concerns deriving especially from the needs of democratic (thus representative and participatory) politics
- it is largely bounded by the limits of the national territory and focuses on 'national interests'
- it legitimates government intervention in communication markets for social purposes
- it generally requires active and continuous policy-making and revision.

While each of these features is important, perhaps the main distinguishing feature has been the similar normative commitments that we find in many different systems, usually expressed in laws and regulations which touch directly or indirectly on *media content*.

5. Changing context for Phase II media policy

5.1 Five trends in communications

Media policy in Phase II had largely ignored the common carrier/conversational means of communication (telephony, mail) until new developments in communication and media technology challenged the regulatory status quo. At present, there are five trends that fundamentally change the context of media policy and put heavy pressure on the 'traditional' way this policy may be implemented. These trends are:

1. digitalisation of information and communication technology, leading amongst other things to convergence between media (broadcasting) and telecommunications;
2. exponential informatisation of society creating an information abundant society;
3. exponential diversification in information and communication products, that is, diversification in contents (tailored information), in information carrying technologies, and in distribution channels and outlets;
4. stagnation in the consumption of information, that is, the demand for information and communication products seriously lags behind supply;
5. segmentation and fragmentation in audiences, that is, the availability of more different communication technologies and more communication channels decreases informational communality and exchange in society.

These trends together result in growing competition between content and communication service providers and contribute to a fundamental change in the structure of media and communication markets. These markets are increasingly demand-driven. The growing gulf between media demand and supply, forces the media industry into a buyer's market. Demand in a buyer's market is economically elastic, that is, reacts immediately and proportionally or more than proportionally to changes in supply and price. In a buyer's market consumers quickly and in great numbers drop out if they have to exert themselves, or if they have to pay a high price for information or if the value of information decreases. As against the elastic demand on a buyer's market, there is often an inelastic supply. That is, a supply of information that is not driven by the price of information, but by technology, production capacity and immaterial goals of the information provider. Moreover, in addition to demand becoming elastic, competition between suppliers for audiences and for consumers of their products will get harsher than ever before.

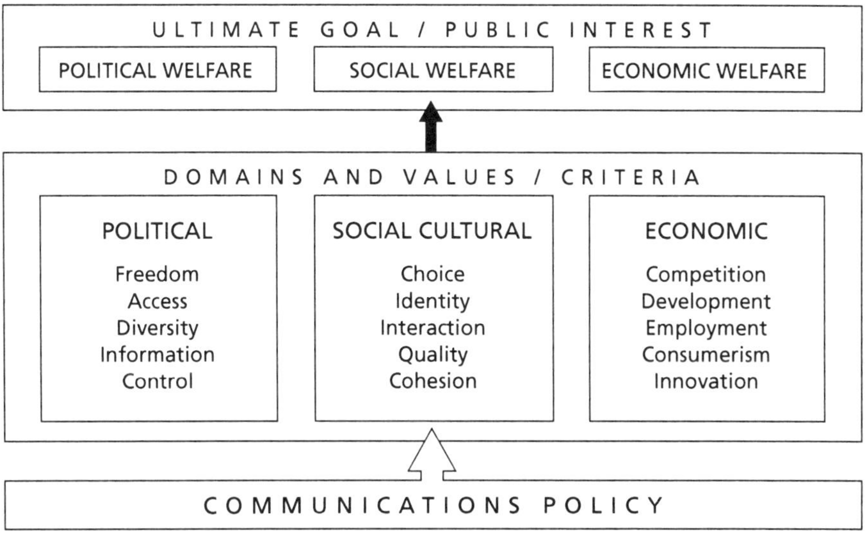

Figure 4: *Phase III* (1980 / 1990 -...):
An emerging new Communications Policy Paradigm

5.2 A note on media diversity and competition

In the Phase II public-service media model, political and social diversity of media content is one of the most important media performance standards. Many theorists suggest that media diversity best flourishes in a free marketplace of ideas: if people can freely enter this marketplace to exchange information and opinions, then we may expect cultural variety and diversity to happen. Of course, this notion is based upon classical economic market theory: if there is full competition in the marketplace, then the market mechanism will most efficiently produce best quality of products against the lowest price possible (Van Cuilenburg, 1998a). Paradoxically, however, increasing competition in media and communication markets may also undermine this basic assumption of Phase II media policy. Media competition not always will be beneficial for diversity of media products. To be more precise, under different conditions media competition favours different types of media diversity. This is due to Hotelling's Law of 'excessive sameness of products. Already in 1929 the economist Harold Hotelling mathematically showed that in competitive markets for producers of goods it is rational to make their products as similar as possible. That's why buyers, according to Hotelling, nearly everywhere are confronted with an 'excessive sameness' of products (Hotelling, 1929: 54). Hotelling's Law applied to communication markets, predicts that highly competitive media markets tend to homogeneity more than monopolistic or public-service media systems (Steiner, 1952: 104-223; Owen and Wildman, 1992: 64-100). For content diversity, Hotelling's Law implies that under different conditions of competition different kinds of media diversity are being favoured: sharp media competition drives out market dynamics, media innovation and openness in diversity in favour of market conformity, middle-of-the-road media products and reflective diversity (see; Van Cuilenburg, 1998b). Against the background of increasing media competition, a fundamental rethinking of media policy paradigms might seem logical and inevitable in the future. Given that market forces under conditions of sharp competition favour media market conformity, one might seriously wonder whether government media policy should not do the opposite, that is, promote media openness and media market dynamics.

6. Challenges to the 'old' Phase II normative policy model

During the 1980s especially, we could see a struggle being played out between the defenders of the dominant socio-political normative model and the advocates of a new approach. In the United States there was upheaval rather than conflict, showing up in the 1984 deregulation which promoted the rise of cable and the 1984 divestiture which broke up the Bell Telephone monopoly and separated it from AT&T in the interests of competition and expansion of telecommunications (Sussman, 1997, 75-6). In Europe there was more of a genuine political debate, although the trend was the same and driven by the same causes. The political aspect showed up mainly in the lead being taken by governments of the right, especially in Britain and Germany, to introduce competition for broadcast television via privately owned cable systems. Britain went a step further than others by privatising and opening to competition the British telecom monopoly. But France had a more or less bipartisan policy of privatisation and state-sponsored expansion of the new media. This reflected a view that one way or another the new technology would bring profound change and also national economic benefits for France and could not simply be held back or ignored (Tunstall & Palmer (eds.), 1991).

6.1 Two policy tendencies

At the outset of this era of European grappling with the challenges of media innovation and expansion, two main policy tendencies could be observed.

One was simply to break monopolies and privatise as much as possible, under the banner of deregulation, along American lines and harmonisation of markets, European style. The second was to keep operating in the spirit of normative theory and try to develop new media potential by way of public investment and protectionism — in effect to apply the public service model to new territory. This required some imaginative leaps, since the democratic gain held out by new media was somewhat uncertain (mainly based on the notion of a more informed, interactive and participant political and social life). The speculative gains were offset by more predictable losses following the decline of public broadcasting and the rise of new media moguls.

As it turned out, both of these strategies were held back by the schizoid character of the enterprise. There was limited political wish or will, and little popular pressure, to discard the existing public sector (broadcasting) or to harm various social and cultural interests, even on the political right. The sacrifice of monopoly was about as far as governments would go to create space for new competitors in the media marketplace. The government — and European Community-sponsored ventures into the new technology were not a great success by any standard, but the causes of failure lay in their prematurity and in the generally lukewarm consumer interest in the promised abundance of information and entertainment. These factors also operated to restrain and slow down the project of private sector-led media growth.

Despite the long reprieve from execution (still suspended), the normative policy paradigm of the postwar era has declined in authority and scope and changed in respect of the means for achieving its goals, even if it retains some legitimacy and popularity. The causes of decline are too well known to need much discussion here. The lever of change has been technology since this made it impossible simply to preserve the status quo. A new socio-political environment (more difficult to pinpoint) has also made it possible to escape from the ideological hold of earlier normative ideas that had their effect through the political system.

The essentially hybrid and interactive nature of new communication technologies has more sharply than ever exposed the lack of logic in the traditional regulatory division into three main sectors, quite apart from the end of spectrum scarcity which provided the main official justification for public monopoly. The convergence of communication infrastructures and services has often been cited as the main cause of paradigm shift, but the most influential causes of change are probably the ambitions of media corporations and governments alike to benefit from the economic opportunities offered by communication technology. This is an additional reason for suggesting that the phase we are entering has more than a few echoes of neglected early history.

7. Phase III: In search of a new Communications Policy Paradigm

Although no distinctive new model of policy has emerged, it is not too early to speak of the postwar normative model as an 'old paradigm', since it has retreated on so many fronts and been extensively adapted. In some respects, the history of Phase I has had to be replayed, but against an entirely different background, with new political ideas and social values. Once again, there are new and powerful technologies with unclear potential for development, vast commercial and industrial interests at stake, governments struggling to keep abreast of change.

Currently, governments have a (national) interest in exploiting the employment and revenue benefits from the expanding markets in hardware and software. Globalisation is helping to drive expansion but has not significantly reduced the role of national communications policy, since national governments, along with national and multinational corporations are still the main actors in the policy arena. Corporations want to develop new international markets and also to expand and merge by crossing old regulatory frontiers. Governments are retreating from regulation where it interferes with market development and giving relatively more priority to economic over social-cultural and political welfare when priorities have to be set. The weakening and transformation of normative commitment can only proceed as far as the political system allows, but there is a good deal of evidence to show that economic pragmatism and re-evaluation of costs and benefits by political actors have given much room for change.

The emerging policy paradigm for media and communications does not lack normative elements but they cover a wider range of values and are less exclusively supported by the normative underpinnings of democratic theory. The relevant norms are in fact noticeably more 'communicative' and less 'political' or 'cultural' in character. This remark depends, of course, on conventional meanings of those terms, in which cultural refers to the traditional art and language of a nation, region or group, while political is defined in terms of the established political institution (especially public communication supports for elections, lawmaking and the practice of government).

7.1 New values in communications policy

The newer communication values, which have been added to the spectrum of relevant norms, include especially:

- openness and transparency concerning private ownership and public control and regulation
- maximum access and participation in communication flows

- maximum choice for users/consumers instead of a guarantee of minimum diversity universally provided
- continual competition and innovation in technology as well as content and communication services
- effective accountability for all public communication
- interactivity rather than just passive reception.

If one chooses to, one can interpret these 'new' values simply as extensions of older ones (McQuail, 1992: 66-73), especially those of freedom and equality. However, there is more than simple extension involved. There are also changes and some older values are losing force. The main area of growing deficit in value terms seems to be in respect of social responsibility, public service and altruism (non-profit goals). The 'public interest' is being significantly redefined to encompass economic and consumerist values (see also model Figure 1).

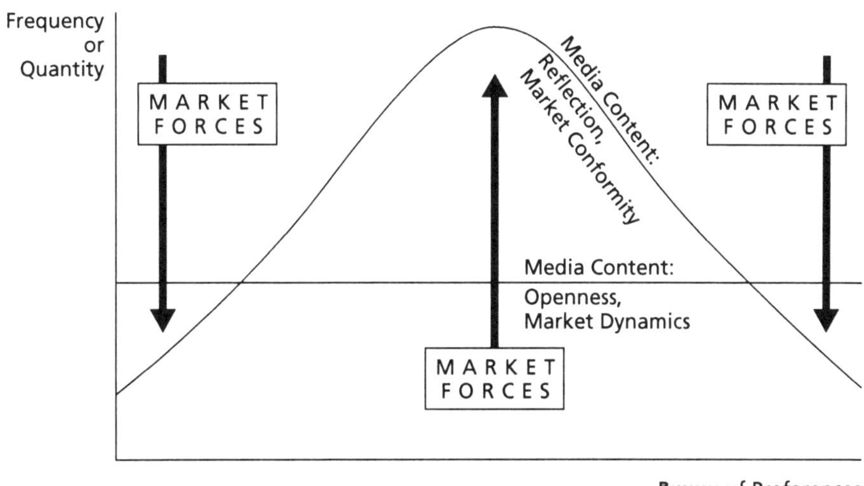

Figure 5: Media Diversity: market conformity *vs* market dynamics

It is not easy to represent the new 'Communications Policy Paradigm' in a single framework, as attempted above, for earlier versions of national communications policy. This is not only because the path of development of communication is still uncertain. It is also because of the continuing contradictions and unresolved dilemmas facing policy-making, not to mention the general uncertainty about the viability of any coherent national communications policy under present conditions. In addition, any new paradigm has to recognise the fact of convergence and probably has to encompass the domain of activity served by telecommunications as well as broadcasting and print. Thus, media policy and telecommunications policy would evolve into one integrated communications policy.

The sketch offered in assumes, as before, a general goal of policy to serve the public interest, but what has mainly changed is the balance of component values that shape the definition of public interest. In general, we can suggest that economic welfare has risen in salience in comparison with political welfare and that social cultural welfare has been redefined with greater reference to the communication values mentioned above.

7.2 Dilemmas and unanswered questions

The Paradigm as sketched is misleadingly neat for the reasons suggested. It is little more than an abbreviated checklist. It is unlikely to reflect the reality of any actual national policy and it is mainly intended to underline the breadth and bifurcation (between economic and non-economic goals) of the current field of policy. The dilemmas and unanswered questions arising for policy-makers include the following:

- whether to promote market dynamics or stability
- what balance to maintain between public and private sector
- how to finance the public sector
- what policy instruments to use for what purpose (as between economic interventions, legal/regulatory moves, communicative means (political influence, information and exhortation, initiatives generated by the system)
- whether and how far to direct policy and control at content or at structure
- if policy is directed at content, whether to focus on reflective media diversity or on open, equal access diversity
- whether to seek a coherent policy for all sectors, or to redraw the map in a new way
- whether to concentrate or disperse regulatory power (and policy making itself)
- whether to maintain national policy or to enter into wider international regulation agreements.

8. Some Concluding Remarks

Until the answer to some of these questions becomes clearer with time it is impossible to be precise about the policy paradigm that applies in a given country. No doubt there will be several variants, depending on such factors as size, the relative share of the information and communications sector in the national economy and the general political climate. It is likely that countries related culturally or economically, for instance by membership of the European Union, will have similar policy projects.

The issue of whether policy will converge, following the logic of technology, by choosing the same goals and applying the same principles and means, within some sort of unified regulatory apparatus (as suggested, e.g. by Collins and Murroni, 1996) will probably not be answered by a simple Yes or No. There are reasons for redrawing boundaries, but no necessity to have only one regime for different kinds of service. There are reasons also for having a coherent set of principles (especially on matters to do with freedom and diversity), but this does not mean that all kinds of content have to be treated equally (advertising, art, news, pornography, etc.). Distinctions of content and audience remain. It is also possible, even necessary, to use different means for different purposes. It is important in the interests of freedom not to concentrate too much power over communication in any single body and there must always be exceptions and alternative routes to follow. Equally important perhaps is the wish to retain a scope for putting into effect political and social cultural values on behalf of those affected by communication, audiences, and citizens.

Freedom of communication, access and control: a concluding plea for a new Communications Policy Paradigm

Just by way of conclusion, we end up with a personal elaboration of a new Communications Policy Paradigm for the future. Though public service broadcasting and universal telecommunication service will most probably not disappear from the political agenda, convergence in technology, liberalisation of communication markets and information and communication abundance, demand a different conceptualisation of communications policy. If our estimate of the future is right, three concepts will be central to any new communications policy model, that is the concept of freedom of communication, access and control (Van Cuilenburg and Verhoest, 1998).

Freedom of communication may be defined as follows:

> Freedom of communication is the right to send or not to send, and to receive or not to receive symbolic messages without hindrance or restraint by any third party.

Freedom of communication will presumably remain the main objective of any communications policy to be designed. In talking about freedom it is common to make a distinction between freedom from ... ('negative freedom) and freedom for ... ('positive freedom'). Freedom from ... refers to the independence of communication participants from government (no censorship), from economic and other forces hindering exchange of messages (e.g. economic monopolies and other dominant market positions) and from social and political pressure groups. Freedom for refers to all those activities people in society can do with their freedom, e.g. to contribute to the civic debate on politics or to express artistic creativity.

Regulating communications may concern different dimensions of the communications system, that is, the structure of the communication market, the market conduct of communication participants or the content of communication (cf. McQuail, 1992: 87-96). Freedom of communication from ... relates mainly to market structure and market conduct, whereas freedom for... primarily relates to content. It is with these three different dimensions (structure, conduct and performance) that the concept of 'access' comes in.

'Access' may be defined as follows:

> Access to communications is the possibility for individuals, groups of individuals, organisations and institutions to share society's communications resources, that is, to participate in the market of distribution services (communications infrastructure and transport), and in the market of content and communication services.

Freedom of communication and free and equal access are related to the question, who is in control of the actual communication process. This brings us to the third key-concept for a New Communications Policy Paradigm: 'control'. This concept may be defined as control over access:

> Control over access to communications is control over deciding who gets access to what communications resources, when, where, how, and on what conditions.

Against this background we finish with a plea for an interpretation of future communications policy paradigms that take as starting point that the main goal of any communications policy should be, free and equal access to a social communications system that diversely provides for the information and communication needs in society. In a context of technological convergence and increasing market competition, communications policies primarily is Communications policy for access, rather than anything else. In focusing on access to all layers of the communications system, communications policy preferably aims at:

- the greatest freedom of communication
- for the greatest number of suppliers and users, and
- should therefore monitor and correct any situation in which control of access is subject to market failure and unequally distributed over political, social and economic groups and organisations in society.

Jean Claude Burgelman

Regulating access in the information society: on the usage and possible reach of present concepts in communications policy and thinking

Introduction

This article focuses on two issues with regard to the policy debate on the regulation of access to communication systems.

The first issue deals with the concepts that are commonly used in communication policy with regard to the regulation of 'access'. In particular, it will be discussed what 'public service' and 'universal service' mean. These are the two most important concepts communication policy has at its disposal in this matter and they will therefore be analysed in their usage.

The second issue of this article raises the vital question with regard to the present policy debates on the validity and usefulness of these concepts in the new communications environments that are being created. Or to put it differently: are 'public service' and 'universal service' as valid as they were and do they have the same usefulness in view of the problems and challenges that new information and communication technologies raise within the perspectives of an evolution to an information society. The latter concept is used in its broadest possible sense; but being aware of its problematic character (Garnham, 1998), this article therefore takes the point of view of the High Level Group of experts from the EU:

> 'The information society is the society currently being put into place, where low-cost information and data storage and transmission technologies are in general use. This generalisation of information and data use is being accompanied by organisational, commercial, social and legal innovations that will profoundly change life both in the world of work and in society generally'. (European Commission, 1997a: 11)

This definition indicates already why, as will be outlined, communication policy in the communications environment created by this new information and communication technologies, needs to be more broadly approached than in the past.

In such a context, it will be demonstrated that the traditional concepts of communication policy reveal their (many different) shortcomings with regard to their applicability, validity and usefulness.

Therefore, this article will end with some reflections on what communication policy in a networked society — or information society — should take into account.

The main argument will be that communication policy has to shift from media policy to social policy. What it means for regulating access and its two main concepts, will also be briefly, in terms of future research perspectives, outlined.

1. Why and how to regulate access?

It is quite largely accepted — as many of the quoted handbooks indicate (McQuail, 1992) that in general a good or balanced communication system, implying good policy with regard to access, must offer: (a) a reflection, throughout its communication system, of the diversity of a given society; (b) the possibility to its citizens to express their values and beliefs; and (c) diversity of choice between media (a real choice between different newspapers, library systems, televisions and so on must be available) and within the system as a whole (the different media have to be there).

Or, as was put forward when the Amsterdam School for Communication research (which concentrates on 'access' as one of its main research topics) was launched:

> 'the main goal of communications policy is free and equal access to a social communications system that diversely provides for the information and communication needs in society'. (van Cuilenburg, 1997: 10)

This typology illustrates that the debate on access is very much linked, at least in Western Europe, to the one on pluralism and that underlying both concepts is the idea that a plural and accessible system, must also be affordable.

In fact, the debate on access evolves around three interrelated sub-areas: how accessible, plural and affordable is a communications system in order to be desirable? The latter being perceived as a more or less balanced state of affairs within each of the three criteria mentioned.

The advantage of such a functional definition of what a good or balanced communication system should be, is that it can easily be used regardless the nature of the media involved. Indeed, it can be applied to the 'old' media system and in fact most of the policy debates are about access to the communications system provided by radio, television, the press, the telephone and sometimes even libraries.

But is can also be applied to newly distributed forms of old media like electronic news media, different types of pay TV and in particular to the 'real' new media, based on effective interactivity, like e-mail, World Wide Web-based applications, which are the result of all kinds of different emerging electronic communications protocols and devices and which are mostly Internet- (or IP- to be more precise) based. This distinction between 'old' 'new' and 'newly distributed', as developed in Burgelman (1995c), is not intended for epistemological purposes, but is only meant, as will be seen, to clarify some points with regard to the consequences on policy.

Regulation of a communications system and regulation refers to that part of the policy activity that has always popped up when at a certain point in time society felt that one or more of these criteria were endangered (For a classic example see De Sola Pool, 1983 or the publications of the Euromedia Research Group. The main driver for regulatory intervention has always been, to a large extent, the feeling that a disturbance in the communications system would result in the fact that the communications system as a whole would no longer be perceived as balanced between and within the three criteria mentioned.

The words 'felt' and 'perceived' are used on purpose here to illustrate that, contrary to the functional definition just given, the regulatory process with regard to accessibility is always subjective because the concepts used are more or less normative.

Indeed, whereas in the 50s most West European societies, on basis of pluralism, would probably never have accepted that less than five communication groups would control the majority of their national media provisions, oligopolistic control of the market is a fact these days. More important, such concentration is, certainly by policymakers, less and less perceived as a problem and if at all mainly seen in terms of anti-competitive behaviour (Pauwels, 1998).

Another example to illustrate the extremely normative dimension of this debate is that after 35 years, the broadcasting policy in most West European countries shifted from the view that a state monopoly was the best guarantee for a balanced and diverse audio-visual provision to the view that only a competitive market, with some marginal corrections, can guarantee this. This debate shifted so dramatically (Burgelman & Pauwels, 1992a) that, whether public service broadcasting could accept publicity or not, it was a heated topic in Western Europe 20 years ago, it is hardly a topic at all nowadays. The debate on the future of public service broadcasting is nothing more than a mere debate as to whether or not a certain country should have (at least) one public service channel no matter how it is financed.

Both examples illustrate that the way the three criteria on communication policy for access are made operational, is very much determined by time and thus dynamic in nature. But also by context and thus subject to certain social, economic and political forces. In the examples just given, it is the changing vision on what should be cultural policy (educating or not) and the role of the state (player or supervisor), as well as the precarious position of state budgets (deficits) that explain the change towards an acceptance of oligopolistic control of the market and niche market role of the public broadcaster.

When analysing the concepts communication policy developed to regulate its defined normative goals, one sees that they were more or less incrementally and self-reflexive. This time and context specificity should be taken into account.

2. The dominant concepts of communication policy with regard to the regulation of access

With regard to the regulatory concepts available to change or correct an imbalance in a communication system, policymakers mostly fell back on the concepts of public service and universal service.

Indeed, both concepts were heavily used by policymakers to enforce (within a certain time and context) a desired type of organisation of their communication system in a more accessible/plural/affordable way. Because commercial broadcasting was perceived as only accessible to commercial interests, most post-war governments enforced a public service monopoly because that was regarded as being the best guarantee for an open and diverse broadcasting system. Because the telephone was only accessible and affordable to the rich and urban elite, most West European governments decided, after the Second World War, to enforce a monopolistic universal service telecommunications system.

Though there are more regulatory concepts (privacy e.g.) and though the arguments are a bit more complicated than can be stated here (Burgelman, 1997b), only these two and their most common explanation are withheld, because one can find them popping up again and again with regard to the present debates related to the information society.

Moreover, they tend to be used in an interchangeable way, which can be explained by the ongoing convergence of transmission networks. Indeed, the type of the network is no longer relevant to the type of service provided, as the case used to be. Broadcast signals can these days easily be transmitted by copper cables and telephone calls can as easily be handled by a coax cable.

It is this convergence that explains why in policy circles, e.g. the EU, every possible problem with regard to access in the future information society is a priori anticipated by the notion of public service and universal service. Should there be a danger that the information society will be a dual one, as deducted from the development towards many Internets? No problem, the universal service fund or universal service obligation will solve it, so that it will be to the benefit of all. Could a problem arise for the diversity of the information society, due to enormous ongoing oligopolisation and sectorial integration of all types of media and hardware sectors? No problem, public service is there to secure and balance high quality content.

More generally, one can say that every problem related to the three parameters of access, policy discourse and research with regard to the IS tends to fall back on concepts of public service and universal service as adequate and universal solutions.

This is a critical point, because as the examples illustrate and as was elaborated at length elsewhere (see Burgelman, 1997a+b), such a use of universal and public service as main regulatory concepts for possible problems with regard to the information infrastructure/networked society/information society is simplistic, formalistic and thus idealistic.

Simplistic, because it assumes that the past functionality of these concepts can be extrapolated to the future. That in other words, the same effects can be obtained now as in the past, by applying the same concepts and their operationalisation. Formalistic, because by doing so, it assumes that these concepts have a dynamic of their own, regardless of the time and context in which they are applied. And idealistic, because it assumes that good laws (measures/intentions/decrees/...) are enough to implement them.

Historical research shows, first of all, that both concepts mean different things: universal service, as applied in telecommunications, is more economic and measurable, whereas public service, as applied in broadcasting, has a more political-cultural dimension and is not measurable. Furthermore universal and public service were the result of some technological facts — scarcity of frequencies in broadcasting or natural monopoly of the network and economies of scale in telecommunications — but they were also the result of political and economical conflict. The application and implementation of both concepts tended to differ amongst nations and cultures as the different ways in which broadcasting and telecommunications took shape in the US and Europe illustrate quite convincingly. And not at least because both came into play in a very elitist environment. Indeed, one tends to forget that when public service and universal service were accepted as regulatory concepts in communication policy, the use of most media or telephone was limited to the elite and the bourgeoisie. The public service provision of broadcasting was mainly a service for the benefit of the well-to-do who could afford to buy a radio-set and for whom the market (public sphere) was their market. Universal service was mainly universal for those who could afford a telephone in the first place. For a long time these were mainly business people.

It is therefore a formalistic error to extrapolate both concepts just like that, to assume (simplistically) that they refer to a perfect policy or to a real 'public' and a real 'universality' that should have existed one day, and then transpose them in an idealistic way to the present day situation which is so radically different and much more complex.

In this respect and with reference to the present debates, it is striking again, as shown in different overviews of communication policy (Schlesinger, Silverstone & Corner, 1997), to notice how simplistically or naively the categories 'market' and 'state' in policy debates on access are used. Most debates and indeed a lot of research, tend to use this as a dichotomic analytical pair: there is a market-way of regulation or a non-market one, inevitably the state. According to the orientation of the author, this is then judged to be good or bad.

Here too, historical research and detailed analysis have shown that it has never been the one nor the other and that, to regulate in function of the market or the state was always but a means, and never an end in itself. Indeed, public broadcasting was for example a compromise between a hesitating electro- technical industry in Europe (to go for private broadcasting) and a dynamic political will to control new means of communication. When, later on, traditional industry went to look for new markets and most governments in Europe seemed to prefer more non critical or impartial broadcasting than the public one, the road was open for liberalisation.

In other words, it is an idealistic assumption that market models and state models exist in their pure form as an end in themselves. Moreover, both have benefits and drawbacks depending on what criteria are chosen to evaluate them (efficiency, innovative power, social dimensions and so on). And again, this differs from nation to nation. And what is certainly clear, both need a lot of a prior thinking and operationalisation. They have indeed no logic of their own.

It follows from these points of view that though the notion of access as such can be used in a 'universal' or diachronic way, just as democracy is universal in a normative way only, the operationalisation of it should be subject to time and context-specific factors.

Which context needs then to be taken into account and how can this serve as a starting point for re-operationalising public and universal service?

3. Communication policy for access in an information society: the changing context

Since universal service and public service are mainly a Western 'invention', it follows that the context to look for is that of the industrial world and, for reasons of clarity, of Western Europe. This implies that the context, with the changing dynamics of the welfare state, is to be seen as the conceptual form within which most Western European countries have organised their societies (including broadcasting and telecommunications) after the second World-War.

Following the analysis of Group of Lisbon (1994), three interrelated changes and dynamics took place in these societies:

i. First of all, there is a change in policy and related thinking in what is considered to be the optimal balance between affluence and welfare.
ii. Secondly, there is a change in the up-till-now traditional values underlying social policy in general.
iii. And thirdly, there is a shift in the organisation of the relationship between private and public time.

To start with, there has been, since at least two decades, a noticeable change in the debate on what used to be considered the optimal relationship between the affluent society and the welfare state. Ever since World War II, the average policy for the construction of the welfare state in Western countries has been based on the existence of a 'social contract' between the individual and the collective, the state. This relationship is considered to be dialectic: the affluent society needs the welfare state just as much as vice versa. A dialectic which has not only created a wealthy but also a stable society, which was judged necessary for an innovative economy axed on the future as well (Ducatel, Fahrenkrog & Cavigan, 1997).

Traditionally this ' social contract' guaranteed:

- the right of labour, by pursuing complete and lifelong employment; by providing collective labour agreements and so on
- the prevention of poverty, by providing minimal income and various types of social aid
- the protection against social risks (social and medical security system)
- the promotion of equal chances by providing cheap, but quality education and culture.

Due to many reasons — the rise of a new conservatism, the pressures of financial capitalism, the economic recession since the seventies' oil crisis and all kinds of technological innovations — an increasing strain has been put on this social contract to a point where it is perceived illegitimate in many people's eyes.

This is easy to notice when considering the following trends:

- full employment is increasingly questioned as a defendable social objective. It is replaced by the idea that we will have 'to learn' to accept 30% of under- or unemployed in Europe
- the prevention of poverty, one of the welfare state's first achievements, is radically hindered by all kinds of cuts or by privatisation (e.g. charity)
- the social security system, which functioned as the welfare society's social safety net and which, together with a guaranteed minimal income, also provided a consumption potential for the lowest incomes, is also threatened by a policy of cuts and decreases

- Though seldom explicitly stated, the promotion of equal chances is seriously threatened, by the upcoming privatisation of education, cultural provisions, media, etc.

For the purpose of this article, it is important to note that communication policy is traditionally situated on the fourth level of the social contract. However, it is clear (see e.g. Mansell & Wehn, 1998, 6-19, 256-263) that in an information society or network-based economy, communication policy will also be vital on the other levels: job creation tends to be more and more information-based (level 1&2); tele-medicine is increasingly gaining ground as a routine in the health sector (level 3) and so on.

The questioning of the re-balancing mechanism that was called the social contract, goes together with the fact that for the first time since World War II, the best regulating mechanism is considered to be a minimally regulated market, in particular with regard to the information society.

Important to note here is that solidarity, an essential assumption in the post-war reconstruction and the metaphor for the collective citizenship model which held good in Western Europe for over three decades, gets increasingly replaced by an individualistic utilitarian citizenship model which raises questions like: 'what is the cost of disease?'; 'how many unemployed can we afford?'; 'aren't pensions a luxury of the past?'; 'should people take care of themselves instead of the state'; in other words the 'what can I get out of it'-attitude (Elchardus, 1994; Galbraith, 1992).

This touches on the second change against which the debate on the future of the welfare state in the information society has to be seen: the shifts in the values underlying social policy in general.

A first shift is the one from cultural paternalism, whereby quality is considered a rare good that has to be offered to as many people as possible, to post modernism where, since quality is whatever the consumer decides it is, is hence not scarce at all.

In broadcasting for example this shift is expressed in the abandoning of public broadcasting in favour of commercial models. Both policy movements, whether or not implicitly, are legitimated with utmost simplicity: the 'state' knows what is good for the viewer vs everything is quality and the viewer is independent in his/her choice.

The second shift tends to be identified as the crisis of Western culture (Morin, 1987). It is a rather philosophical debate pointing to the feeling of large parts of the Western population that their societies no longer follow a 'grand social' project and that as a consequence meaningful world-views guiding such a project are missing. Which might explain the increasing reliance on thorough and 'proven' values and norms like language, race, religion and other forms of cultural fundamentalism. Not to mention the upcoming of superstition: the pyramids as healers, talking to the dead, the Genesis as the explanation for the creation of life (Levy, 1994).

Finally, when thinking about the welfare state in the context of an information society, a third global change has to be mentioned: the shift in the organisation of time and space, more precisely in the relationship between private and public time (or life) (Ariès & Duby, 1990: 13-126).

Though a lot can be said about the banalisation and transformation of time and space, an example can be given by looking at labour and how information and communication technologies change the way labour is organised in time and space.

The history of labour and the organisation of it, is characterised by a permanent evolution to distance itself completely from the private (home) and into the public domain, where it got organised according to standardised and usually highly formal procedures (nine to five, etc.). The whole societal system of social organisations such as labour unions, political constructs like collective bargaining of labour conditions, but also the whole way our education system is conceived, is built on this.

ICT can simply turn back the clock and bring labour again in the private sphere (e.g. telework). Very appealing but with enormous side effects on the social, political and other organisational constructions mentioned above. The school system, to name one, does follow the hours of labour and one of the main problems of labour flexibility as a result of introducing ICT is e.g. the fact that families have difficulties organising their time in accordance with that of the schools (getting the children from school e.g.).

How do these trends fit together with regard to the debate on communication policy?

On the one hand, it is clear that not only the traditional parameters of the social contract are under pressure, but also the assumptions (social cohesion etc.) on which it was based. Moreover, this coincides with a quite dramatic change, made possible by new information and communication technologies, in the organisation of our time and space and the societal institutions that are based on it.

On the other hand, the communications sector has also an enormous impact on and hence is essential to at least three of the four parameters of the social contract (where it used to have an impact only on one level). The communication sector is thus no longer an ephemeral context in which, since World War II, the welfare state has been organised, but it is central to its further development. The debate on the regulation of access should be seen in relation to this radically changing societal context.

4. To what degree are the existing regulatory mechanisms/concepts of communication policy efficient in view of future developments?

From the changing environment just outlined follows that universal service and public service should be fundamentally re-defined (and not just re-applied) and re-operationalised (and not just extrapolated).

On the one hand (with regard to a redefinition) this is so because both concepts focus only on one level of the social contract (level four) and the 'revolution we are living' covers all four. When discussing tele-medicine e.g., there is clearly an aspect of access related to it. That is, if policy wants to avoid a dual medical or social security system. However, universal service in its traditional definition does not cover that at all.

On the other hand (with regard to a new operationalisation), both concepts are sector-specific and apply exclusively to only the broadcasting and telecommunications sector, as stated above.

As a result, the enforcement and operationalisation of these regulatory concepts have been sector-specific up till now. In the case of the public service, this was done by a special tax, a license fee for broadcasting whereas in the case of universal service the installation of monopolies and the allowance of cross subsidisation were — for a long time — the preferred enforcement mechanisms.

The question remains however whether communication policy can still limit itself to sector specificity, given the changes outlined. Or to put it rhetorically, is it enough, in an emerging communication world characterised by abundance of provisions (which is not the same as quality) and segmentation of use, to say that a public service policy consists in offering one public service channel? And is it enough, in a world where access to Internet, or more in general to a global network, will be so vital, to limit universal service policy to the provision of a phone within a reach of three kilometres (which is the ITU definition of access)?

5. Policing access in the information society: towards an integrated and cross-sectorial approach

Clearly, in an information society scenario communication policy cannot limit itself to the traditional and classical issues of diversity in media and equal access to telephone. Nor can it be sufficient to restrict communication policy to the provision of access to channels or media alone. If that were so, policy would be irrelevant since competition as such provides for more and more media channels and cheaper and cheaper access. This is why the dominant policy discourse, by only focussing on ICT as media channels, is intrinsically correct: the more competition, the bigger the offer (satellite television), the cheaper the transmission (Internet) and so on.

Only if related to the question of what content of what quality is available on these cheaper premises, and to the question of which capacities are needed to work with this content, the dominant discourse reveals to be problematic.

But rethinking this should go much further. Since the communication sector itself is becoming the ground on which the four levels of the social contract will increasingly be elaborated, — and hence the ground on which the debates on labour, education and democracy will develop — communication policy cannot be but considered cross-sectorial and has to be seen as part of an overall social policy.

How far this rethinking can or must go and how this relates to economic policy, welfare policy and even international relations, is an enormous research challenge. Recent work (Blackman, 1998; Mansell & Wehn, 1998) is beginning to address this, but it must be clear that the debate is only starting (see several contributions in Calabrese & Burgelman, 1998).

Related to that is — if this cross-sectoriality is accepted — the question of how to operationalise and enforce such a 'new' cross-sectorial communication policy.

Two random examples can illustrate this:

> Applying and enforcing universal service or access in an Internet environment — a defensible option from this point of view — only via the sector itself (e.g. via a universal service fund) will in fact only favour the incumbent telecommunication operators and hence make fair competition problematic (Mansell, 1993). It is quite clear that if tele-medicine is important, the access to it cannot be paid by the operators as such. However, someone will have to pay for the very high expenses inherent to the technically demanding networks a tele-medecine environment needs. This means that the debate on communication policy with regard to tele-medicine should also be a debate on redirecting social and medical policy.
>
> Applying and enforcing public service in a world of networked abundance clearly means being active beyond the classical distribution platforms like cable or satellite, but also on the net as such. Given the upcoming importance of push technologies, which are in fact a networked model of broadcasting, the successful ones at least going in that direction (Uzan & Fontaine, 1998), it seems legitimate to raise the question why public service should not be active here. Or, to push the line even a bit further and look at the trend whereby in the Western world the access cost to the network as such will probably become very reasonable and is forecasted to go down to 150 ECU a year (Dimech & Turpin, 1998), one can argue that the main policy problem here will shift towards the question of access to a quality network or quality content. Here again, the

question is valid if there is any reason, considering the history of public service in libraries, why a public service policy shouldn't be considered with regard to web sites, search engines or portal sites too?

However, if action on this level should be considered, it is clear that financing this only by subscription or yet another fee, the main operational mechanism up till now, is absurd and probably counterproductive, also with regard to the existing public channels. Therefore, looking for cross-sectorial ways of financing a 'new' communication policy, is a second research challenge.

Conclusion

This article starts from the observation that the transformation towards an information society is not only transforming the media or creating new distributed media, but is also transforming firms, economics, the way people live and even governments.

Therefore the impact of the transformations that are going on will go far beyond the traditional domains of communication policy. In fact it will also have a fundamental impact on the global policy framework (social contract) within which most West European countries were constructed since World War II. And though this was not the focus of this article, it is quite clear that the same analytical reasoning could be applied to other continents or even the world. At least this is suggesed by the holistic approach — communication policy as a mix of infrastructure, capabilities and skills — offered by Mansell & Wehn (1998).

This implies that since communication policy has up till now traditionally been sector-specific and mainly devoted to access in media matters or to the telephone network, regulating access in a more networked context should be pushed into new directions.

Not taking the outlined contextual shifts into account means that regulation of access could indeed be limited to supervising fair competition and favouring marginal market intervention. If the source of concern is only the media and/or access to the phone (or the net), then indeed a minimalist regulatory approach may be enough (Van Cuilenburg & Verhoest, 1998 or the various OECD policy documents with regard to telecommunications and media policy, including electronic commerce) simply because the policy goals are minimally defined.

If however the contextual shifts outlined above are taken into account, then it unavoidably implies a radical rethinking and — equally important — a re-operationalisation of the concepts of access and even more so in general communication policy as such.

Here lies a challenge for future research.

Marc Bogdanowicz
Marie d'Udekem-Gevers

Social Regulation: Synthesis of the Debate

'Setting Welfare Objectives for Communication Policy'

1. Setting objectives

As chairman of the workshop on Social Regulation, Prof. Jan Servaes started off by pointing out that all the presentations of the morning's keynote speakers were announcing a shift to more socially-focused concerns in communication policy. Nevertheless, Prof. Servaes also pointed out that these concerns were still expressed at a rather general and abstract level.

As an initial answer to this remark, Professor Jean-Claude Burgelman's presentation offered a first framework for clarifying the objectives of such social concerns. From his point of view, the social contract, defined as a series of guarantees concerning the right of labour, protection of poverty, prevention against social risks and promotion of equal chances, is currently threatened in all its aspects by the rise of the Information Society.

Focusing on media policy, Prof. Jan van Cuilenburg also confirmed in his presentation that from the 80s onwards, technological, economic and social trends have fundamentally changed the context of media policy and that a new communication policy paradigm has to be found. This new paradigm would have to answer such questions as the future of the public sector and its core missions, and the legitimacy and means of content control, if any. Again, political, social and economic welfare was presented as framing necessary ultimate goals.

The Group of Lisbon's reflections (1995) offer another framing tool for determining the objectives of a redefined policy framework. As such, this frame is an urgent plea for a set of four worlwide contracts aiming at equal chances for accessing basic needs, cultural dialogue, world democracy and sustainable development. Its specificity, if compared to other frameworks, is to broaden the objectives of regulation to encompass issues such as democracy and environmental issues, considered as systemically linked in a globalised economy. It also competes with the opinion that communication policy would be exclusive or central to the issue of future welfare in a so-called Information Society.

The nuances between these three complementary frameworks were little discussed. Nevertheless, there was a major consensus among the members of the workshop when considering the necessity of:

- a broad definition of the (re-)regulation's objectives, based on a reflection on the core issues that are at the root of a contemporary social contract (as opposed to narrow adaptations of sector-focused regulations)
- a consequently creative (re-)regulation of communication policy (as opposed to constraining regulation to the undesirable effects of a strictly competition-regulated society and/or market).

2. Making objectives operational

A far less consensual issue concerned the role of the State and the means that have to be mobilised to generate and control new communication policies in a globalised world.

Social Equity and basic needs

When considering universal service as a means of fulfilling the 'Equal chances for accessing basic needs' objective, the concept of 'accessibility' was questioned.

Professor Van Cuilenburg argued that: 'communication policy for access should aim at the greatest freedom for communication, for the greatest number of suppliers and users, and should monitor and correct any situation in which control over access is subject to market failure and is unequally distributed between groups in society'.

Participants in discussing this proposition clarified various conceptual and operational aspects related to this 'accessibility' issue by:

- opposing 'access for use', as a guarantee for free or facilitated access e.g. to equipment and services in the area of tele-education or tele-medecine, to 'access to a content' which refers to a cultural objective (see below)
- underlining the necessity of organising accompanying educational measures to guarantee the effectiveness of access
- distinguishing between awareness-oriented vs skill-oriented training, between fostering effective participation of all vs 'mere' workforce adaptation to new production processes. This emphasis on training policy echoes the Commission's Green Paper (European Commission, 1996f: 16)

144

- imagining public support services such as community-based networks and other enabling services and «empowerment» initiatives
- questioning the realism of the proposals when considering the effective state of basic needs in the world, their affordability when observing the state of public finances, and their necessity when considering that they reinforce a fast-growing industrial market. The discussion then turned towards the funding of these needs
- evoking the public or private financing of the information society. A debate about specific or general taxing systems occurred, as possible determinant for supporting equity- or 'happy few-'oriented trends
- discussing the role and reliability of the State regarding such issues. As quoted also by the Group of Lisbon (Groupe de Lisbonne, 1995: 27), participants coming from countries with (ex-)authoritative regimes testify to the unreliability of the state system in specific circumstances to guarantee basic needs for everybody. Nevertheless, the term of 'Enabling State' was proposed to characterise this role for national public authorities.

Earlier suggestions of the High Level Expert Group on the Information Society appear to be in line with this discussion, notably when it claims in its intermediary report the need of shifting to a concept of universal community service as opposed to a universal individual service;

> 'There is a need to investigate whether, in order to avoid exclusion (...), the existing concept of universal service should not be shifted in the direction of a concept of universal community service, extending universal service provision to incorporate a basic level of access to new information services, but limited in its obligation of universality to the educational, cultural, medical, social and economic institutions of local communities... It would (...) involve, where necessary, public funding for technical and financial assistance' (European Commission, 1997a: 41-42).

Cultural diversity and dialogue

If aiming at a 'cultural dialogue', should a public authority provide itself with any specific content? In Professor Burgelman's opinion a government should only be an 'enabler' in this domain and should not be a provider of content. This opinion directly questions the very existence of public services offering information or entertainment, such as state-financed national broadcasting.

On this topic, opposite positions in various areas were discussed:

- if public broadcasting was traditionally considered as the better guarantee for cultural diversity, the risk of governmental ideological propaganda or cultural imposition is today heavily emphasised

- to oppose the cultural paternalist argument 'Who decides what is good for people?', a new plea was made in favour of schooling and formal education as a basic tool for socialisation and cultural empowerment
- State censorship power was discussed as a means or a threat, including the legitimacy and effectiveness of filtering tools to be delivered to the citizen
- demand-oriented, that is market-driven, diversity was criticised. On this basis, Professor van Cuilenburg's plea for a 'new communication policy paradigm focusing on media market structures rather than on media content' was questioned.

Clearly enough, the regulation of content appeared a much harsher issue among the participants to the workshop. Cultural diversity, and furthermore the means to encourage dialogue between cultures raise more questions than it provides concrete answers.

Democracy at all levels

This objective was not specifically put forward. Nevertheless some arguments have been developed which might lead to an overall understanding of the issue:

- potential applications to foster democracy such as the possibility of public-funded community networks, were presented
- the threat of out-of-control communication tools, both for the state as for the citizen was put forward
- the question was raised of determining the adequate national and international institutions to address all these issues in a democratic way. It underlines the existing contradiction of managing globalisation within subsidiarity, societal issues within media policies, the Information Society project within a group of industrialists or civil servants of the EC. Part of these questions are also addressed in the Group of Lisbon's report. It was suggested to consider the existing regional, national and international institutions and to offer support to open up the debate on communication or media policy within those institutions.

Sustainable development

The link between sustainability and the Information Society had been largely put forward and analysed within existing institutions such as the United Nations (Mansell & Wehn, 1998). Little has been discussed on this topic during the workshop.

3. Conclusion

It may be concluded from this discussion that communication policy, and thus regulatory principles, are much more difficult to set once they are bound to an international context. Managing cultural differences and encouraging dialogue and diversity, identifying or initiating democratic procedures, developing global sustainability, all these objectives necessitate a substantial knowledge of the international scene and cross-sector issues. This expertise is, for obvious historical reasons, not ready-made taking our national regulatory habits and their analysis into account.

PART III

Market Regulation

Eric Van Heesvelde, BIPT

Introduction

Market Regulation

Little more than a year ago the Belgian legislator liberalised – at the very last moment – the telecommunications sector in Belgium. This may well be the time to make a first evaluation.

Such an evaluation is urgently required as it is immediately obvious that liberalisation has not given the residential user what he might well have expected. It may be true that an unmistakably favourable evolution can be seen in the GSM market, with regard to tariffs and service offer, entailing a spectacular growth in penetration, but as for classical telephony and internet rates one can hardly speak of a huge success for this group of users. The tariff reduction for international and long-distance calls is only a scant consolation. In view of secondary legislation, which still needs to be organised, it is necessary to start making a number of analyses now, so that these can already be taken into account. In the meantime, basic legislation has been completed, co-ordinated and adapted to a number (not all) of European criticisms: but here too a lot work remains to be done.

One may wonder if the market is functioning properly. It has been suggested above that this is not quite the case. On the other hand it cannot be denied that a number of national and international companies are showing interest in the Belgian market. Forty licences have already been granted; twenty for infrastructure, twenty for voice telephony, twelve operators have requested a licence for both. Seven licences are currently being examined by the BIPT, and six hundred declarations of the exploitation of all kinds of services have already been registered.

However, there is a but. Not everything is running smoothly: the sector (and therefore also the BIPT) has faced difficulties caused by the not yet complete and incorrect transposition of the European directives. Moreover, all parties involved in the sector, competitors, Belgacom[1] and the BIPT go through a learning process, which is not always easy. That learning process is concerned with formal and informal arbitration by the BIPT, the resolution of delicate dossiers in practice (this will be discussed later on, but reference can also be made to the national and international pressure regarding the unbundling of the local loop, the complex matter of allocation of number capacity, where there already is a shortage in a number of sub-domains), the need for fine-tuning regulation and the question of the relationship between a sector-specific approach and general rules regarding fair competition.

The question also remains whether it still makes sense to hold on to the difference between declarations (of services) and licences (for networks and telephony). Because of market developments it has little sense to maintain this difference. The growth and the economic and social importance of data communications, multimedia services, electronic commerce (generally speaking Internet), is such that there is no use whatsoever to subjecting voice telephony to a more strict regime than those services. As a matter of fact infrastructure and telephone operators now have to work in a much stricter framework than other service providers. The BIPT has few or no resources to act in a strictly regulatory way concerning the declaration of services: it cannot suspend or refuse the exploitation of such services – unless there are serious and manifest infringements on the law, not for instance because the service involved disturbs the market - the BIPT can not even submit the exploitation of such services to certain conditions. It is therefore impossible to anticipatively force the dominant operator (and in the future operators with significant market power) to run its services while respecting the basic rules concerning fair competition. This refers mainly to the provision to competitors of the facilities asked for under correct conditions, so that they can also provide such a service within a reasonable time period. Duet and the provision of internet services through non-geographical numbers, both services being provided by Belgacom, are typical examples of this.

A number of experiences show the gaps in the legislation, or the complexity, so that the regulator should be given more to intervene competencies, or preciser ones. Those experiences concern interconnection, duet, internet, numbering. We shall elaborate on these examples a little further:

> As for interconnection the legislation does not contain any precise instructions for the BIPT. This means that the discretionary power of the BIPT, but also its administrative liability, is very large. The legislation does not say what cost methodology the BIPT has to apply (the range is broad, starting from the fully distributed cost systems to the long-run incremental cost systems), which has far-reaching consequences for the competitors of Belgacom because the tariff level can differ substantially. Neither are qualitative aspects of this reference interconnection offer described in the law: here too the BIPT has discretionary power, e.g. to intervene in service level agreements, to introduce zonal interconnection tariffs and carrier access codes, to schedule the concrete realisation of

interconnection and the course of the negotiations, the obligations between operators regarding the reasonable character of reconfiguration of the networks, the mentioning or not of non-geographical numbers in the interconnection offer. During these first two years of the liberalisation of the telecom sector (1998-1999) the BIPT has always chosen a careful approach in all these matters, as well as for tariffs. It is however quite obvious that if Belgium wants to connect with the countries performing best regarding telecommunications services, definite endeavours should be made before the year 2000 in order to join the European benchmarking, not only concerning tariffs, but also regarding service. Another problem which has revealed itself during the past two years (but which will probably be solved soon) is the lack of a royal decree organising the procedure for the Chamber of Interconnection. As is well-known, this Chamber of Interconnection has the competence to settle disputes between operators during the realisation of an interconnection agreement. It is completely independent within the BIPT.

Duet and Internet. Although these are two totally different services, they have clearly shown that the BIPT needs a reinforcement of its competencies. The first, Duet, is a service that Belgacom wanted to offer in co-operation with Proximus[2]. The service can be described as a high-performance but very complex system of switching through to either the fixed or the mobile network, in connection with a voice mail system. The second is a Belgacom proposal to offer internet through non-geographical numbers at special rates for end users. In both cases the BIPT could do nothing but registrate these declarations of Belgacom strictly according to the legislation, albeit under the restriction that Belgacom had to respect the rules regarding fair competition, including the ban on dumping and cross subsidisation. This meant that Belgacom had to offer similar services under similar conditions to its competitors and to competing internet service providers. Although the latter guarantees had not been fully realised, the BIPT was unable to stop the declaration and as a consequence Belgacom was able to start up both services. The BIPT tried to rectify the situation by forcing Belgacom to negotiate duly with its competitors, basing itself on among other things the legislation on fair competition and the general principles laid down in the telecommunications Act of 1997. From these cases can be concluded that the Institute should be given the competence to suspend, in a number of cases, the exploitation of certain services that have to be declared until certain conditions have been imposed on the incumbent operator and/or on operators with significant market power.

Numbering. A number of similar remarks also apply here. The regulatory framework is very general and it is up to the BIPT to make it concrete. Every time when operators offer a service they need special numbering capacity for, the Institute has to take a decision. When doing so it has to rely on general principles, such as those laid down in the legislation and the relevant royal decree. Because of these interpretation margins the Institute must be very careful on this matter. Another problem arises in relation with transparency: on the one hand the allocation of new number capacity for new applications is very confidential with regard to the operator making the request, on the other hand it creates precedents for new services which can also be claimed by other operators. The BIPT tries to solve this by means of circular letters that lay down the general guidelines. Additional legislative measures will have to be taken in the short term in order to adapt Belgian legislation regarding number portability to the relevant European rules: the introduction of portability on 1 January 2000, the introduction of non-geographical number portability, the fixing of the cost price for this portability. For the latter problem the BIPT is now consulting both the operators involved and the regulatory bodies of the Member States.

All of this concerns the need for a change of the legislation based on the experiences of the past months. The telecommunications sector is however constantly changing. Both technical and economic developments, integration beween operators (horizontally and vertically), the creation of new needs, all require prospective work in order to anticipate these evolutions. The big themes addressed here are the third generation of mobile communications (UMTS), the convergence of fixed and mobile communication, the unbundling of the local loop, voice over internet, convergence towards multimedia, internationalisation of electronic commerce and in general the growth of services vs voice. It is evident that all these developments have such a big impact, both on the sector and on society in general, that the regulation will have to be adapted on several levels. Precisely because of the international context this will have to be done not exclusively on a national level, not even on the level of the European Union, but to a large extent on a global level.

Despite the need for an adaptation of the regulatory framework and in order to eliminate the existing shortcomings on the one hand and on the other to anticipate new developments, a number of rules of thumb need to be applied. The first one is the need to stick to regulation aimed at the sector for yet a long time to come (probably longer than many are currently claiming) before switching to the follow-up of market developments in the sector by way of general rules of fair competition. Apart from that the regulator will have to be reinforced in this period, with regard to manpower and competencies (as shown above). It is clear that the political authorities should draw the general lines on the legislative level and fix a policy regarding fair competition, guarantees for the universal and the public service, industrial and technological evolution, evolution of internet and electronic commerce. Within that framework the BIPT can then play its role in full autonomy.

Finally, on the agenda remain a number of specific items:

- the legislation needs to be more transparent and its adaptations must not endanger the continuity of the rules, since companies need legal security
- the relation with other organisms must be defined: both on international level and towards the Belgian regions and communities and to the council of competition
- there is a need for a fixed and clear legal framework in which the creation of precedents by the BIPT, through explanatory notes, advice and dispute settlement has a sound legal basis.

Notes

[1] The old PTO and thus dominant actor in the Belgian telecommunciations market
[2] The GSM operator of Belgacom

Jean-Marie Cheffert

The Universal Service: a conceptual approach

Introduction

In recent years the telecommunications industry has been subjected to various evolutive forces, the greatest of which has been its opening up to competition. The idea of a universal service has played an increasing part in the political debate these changes have given rise to. However, the connotations arising from this concept are very diverse. At times it is seen as the ally of a very 'economistic' vision of deregulation, at other times it is seen as a voluntarist policy, in opposition to market forces, bent towards the installation of a so-called information society.

Beyond such simplifications, it is important to seize the substantive marrow of the universal service idea and its real connection to the liberalisation of the telecommunications sector.

This conceptual effort aims at bringing back the idea to its just focus: we shall observe that the margin for political manoeuvre is very large and compatible with very different judgements regarding the legal foundation for and nature of public intervention. The scope of the universal service remains basically the prerogative of states. In this regard we shall show how European directives limit their freedom of action (less than might seem to be the case).

But the universal service's scope is not the only object of the political debate. All too often, choices bearing on the *organisation* of public intervention are relegated to the back burner. They are decreed as being dependent solely on legal technicalities or the introduction of strict notions of accountability. Such an attitude reveals a lack of political culture. We consider it crucial that the actual logic of a universal service, its different areas of choice and their interdependence, should be well understood. Thanks to that effort of understanding, society as a whole may come to see its objectives better realised. Objectives which may arise from very different concerns, such as the struggle against a two-track society or a growth in market penetration.

Meeting these goals, freely chosen by society, calls for the construction of a system the logic of which we must first grasp. Understanding this will enable decision-makers to extract genuinely informed choices. This political culture will clothe with meaning those ideas that are often presented as techniques but are in fact based on choices.

The present document aims to examine the universal service both its various component parts as well as its objectives, its domain of action and its relations with the competitive environment. A certain lack of focus, upon which several accepted notions compete, surrounds this idea. We are of the opinion that an initial conceptual effort is called for which will enable us to offer a definition of the universal service idea based on its intrinsic characteristics and independent of any particular sector, such as telecommunications. To this end, we shall draw up a unified logic framework in which different interpretations can find a place. The telecommunications sector, which interests us particularly, will be used to illustrate the examples.

This first reflection, of a conceptual nature, will be made under the first three items of this article. We shall show how the universal service may be seen as a particularly political response to certain goals of public interest. It is important that we grasp the logic of this response since it will place its constraints upon any political action.

The culture of political choices we shall be proposing can be viewed as a step which may lead to the development of an analytical grid. This, once refined, will in turn permit us to revise our reading of European and national choices and evaluate their relevance.

1. The Universal Service Concept, An Initial Approach

In order to grasp the universal service idea, we shall initially deal with European texts applying to telecommunications. Considering both the logical coherence of this approach and its gaps, we shall attempt to delineate the precise contours of the universal service idea and the inherent particularities of its political method. Even though the telecommunications sector serves as model here, our ultimate aim is to approach the idea independent of any specific segment to which it may be applied.

1.1. Goals and scope of a universal service

According to the European approach the universal service is a set of minimal services of a given quality, which are accessible to all users independent of their geographical location and, in the light of specific national considerations, at an affordable price (European Commission, 1998c: article 2f).

This first approach envisages a list of services for each element, meeting three goals or demands: quality, accessibility and affordability. We shall be making further use of this expression in the pages that follow. This approach, rather than 'definition', should be viewed as a logical framework (a list of services to be provided with objectives whose content remains to be decided) enabling us to precisely define the universal service's scope.

We define this scope then as much by the list of services as by the precise contents of the objectives. This logical framework, presented in figure 1 below, may be applied to every economic segment. Filling out this framework serves to define the universal service's scope. The free of charge nature of emergency calls signifies, for example, that this 'item' is in the list and that its quality, its accessibility (everyone) and its affordability (free of charge) have been targeted.

Goods and services affected	Objectives targeted		
	QUALITY	ACCESSIBILITY	AFFORDABILITY
Private telephone Public telephone ...			

Figure 1: A logic framework to define the scope

As regards the telecommunications sector, the national legislator may, in order to define this scope, invoke to a certain degree the subsidiarity principle. Certain European imperatives have a bearing, as we shall see, either on the content of the list or on the content of the objectives targeted (European Commission, 1998c: chapter II — point 4).

Specifically, the legislator may make the telephone service an element of the list by defining its quality and by making it accessible throughout the country at a price whose evolution, for example, may not exceed the rate of inflation.

Up to this point, nothing has been said of the means required to achieve these ends. This approach to the universal service (a logic framework designed to delineate its scope) neglects certain facets of the idea that we must also see as a particular method (or political procedure).

Our exact intention is to approach the universal service in its largest possible sense, that is to say, in its political aspects. At the beginning of this article, under point 3, we presented such a definition. To better outline its contours, we must first consider as much the context in which it is drawn up — which, we shall see, is one of competition — as the method used to achieve the goals of quality, accessibility and affordability. The issue is one, as shown below, of a series of policies (of designation, remuneration, ...) carried out in conformity with a competitive context.

1.2. The functional universal service

The three objectives presented above are compatible with different policies such as, notably, the organisation and regulation of public monopolies. The goals of quality, accessibility and affordability were, in the case of voice telephony, those envisaged globally by a 'public service' policy. However, the objectives were targeted at a particular context, that of monopolies, and according to a particular method, drawn up by means of legal texts, legal or contractual in nature, designed to organise and regulate these public bodies. The organisation of public monopolies may therefore be considered as a specific political response to the three goals given above.

Put otherwise, the stated ends of the universal service, the goals of quality, accessibility, and affordability, are, globally speaking, those of a public service taken in its functional sense (Flammé, 1989). One may, in fact, through a re-reading of the principles of equality and continuity governing the concept of a functional public service, approach them to the goals of quality, affordability and accessibility. This goes beyond the aims of this present article. As a preliminary approximation, we can consider equality as pre-empting price, quality and access. Similarly, the targeted continuity is that of a service whose quality, affordability and accessibility are given values. According to the idea of a functional public service, the three objectives should be achieved in a continuous manner and in respect of equality. This is the case with the functional universal service. The 'law of mutability' must be drawn closer to the evolutive nature of the service's scope (cf. point 1.4). We propose, in a symmetrical manner, the idea of the functional universal service, that is to say, understood in terms of its missions and their stated ends.

We therefore define the idea of the functional universal service as being any policy targeting the three objectives cited above. If we can accept that the ideas of universal service and public service, taken in this functional sense, intersect, it is important to note that the methods used to achieve these same three objectives may differ widely.

1.3. The conceptual universal service: a particular political response

If the notion of a universal service (we are here speaking of the universal service in its currently accepted form, that is to say, including its political aspects: it cannot be merely reduced to its aims or scope) in Europe has currently taken over the baton from the idea of a public service, this is precisely because a change of context, i.e. the opening up to competition in the telecommunications segment, has logically necessitated a different political response to fundamentally unaltered general goals, even if the scope, as we shall see, is an evolving one.

The 'conceptual' universal service goes far beyond its objectives and its field: being characterised specifically by a context, a competitive environment, and a policy particularly linked to this context. We shall demonstrate how competition is one way, among others, of achieving the targeted goals.

1.4. Two reductional views

1.4.1. The universal service as a scope

It is important, at this stage, to air certain confusions or reductions. The first consists in reducing the idea of the universal service to its scope or domain of application. The universal service 'will be' this or that service to meet this or that demand made upon it.

According to this logic, there is a tendency to oppose the ideas of public service and universal service on the basis of their domains only. While the 'old' notion of public service only concerned the telephone, the more modern idea of a universal service embraces the entire information society. It envisages, in time, to place a demand on other services such as the Internet, video-conferencing in schools or hospitals, the availability of public information, ... As we have seen, the novelty lies elsewhere, in the change of context and therefore the method of response. It is nonetheless true, though this is a parallel phenomenon, that the scope of a universal service is expected to extend or alter itself, as was formerly the case with the functional public service, according to the 'law of mutability' applicable (Flammé, 1989).

Such a reduction of the universal service concept sometimes tends to truncate the political debate, which often only bears on the scope itself, thus neglecting the political choices to be made at different stages of the system's development (universal service supply policy, finance policy...). These elements will be presented under point four.

In our opinion, a simple debate on extending the scope makes little sense if the system currently in place is lame. Such is the case when the slightest enlargement of the scope involves expensive changes to the legal framework. In other words, this essentially evolving debate on scope enlargement, if it is to avoid becoming vacuous, must involve a debate on the system itself. This presupposes an understanding of the latter's logic and interdependencies, which is precisely what the present contribution is aiming to provide.

1.4.2. The universal service and the operator's status

If the universal service is sometimes reduced to its scope alone, there is yet another confusion which sometimes falsely refracts the political debate. This consists in associating the universal service with the private status of its operators (and the public service with the operator's public one). According to this view, the introduction of a universal service will initiate the privatisation of its traditional operators. Nothing could be further from the truth. Both at the level of logic and policy, the concept of a universal service takes up, as we have seen, the baton of the public service as soon as this is opened up to competition.[1] Universal service policy can be equally well applied to private or public operators, as is the case in France.

To illustrate this, we should note that where the regulator obliges this or that public or private operator to provide non-profitable services in a competitive environment, the operator must be remunerated according to a method coherent with that environment. Non-profitable services would be the kind of services that are not spontaneously offered. In particular, if an unfair distortion of the competition is to be avoided, such remuneration should place him, in fine, in a comparable financial situation to that of an operator without any such obligation. Universal service policy can therefore be seen as disposing of a logic that transcends the private or public statute of its operators.

It is clear that the idea of the conceptual universal service represents a particular political response, made necessary by a change of context: the opening up to competition. It is this which has enabled the idea and its logic to germinate, not other phenomena observed in parallel, such as the privatisation of public operators or the expansion of the scope of public intervention as the result of the advent of the information society.

2. Universal Service Policy, Its Context And Logic

In this section we shall analyse both the links that the universal service maintains with the competitive environment and the intrinsic logic of its method. This reflection, illustrated by the telecommunications sector, is intended as general. It will lead us to a definition under point 3.

2.1. The importance of the competitive context

Following a political debate, the decision was taken in Europe to open up the telecommunications sector to competition. This competitive context is a crucial element in the analysis of the universal service concept for at least three reasons. These go far beyond the boundaries of the sector which concerns us.

Firstly, (a) the competitive context is, in itself, expected to participate in the attainment of universal service objectives: quality, accessibility, affordability. One expects, in a global sense, that competition will cause a dramatic fall in prices and an increase in the availability of products. Such expectations argue in favour of the opening of the telecommunications sector, as others, to competition.

Secondly, (b) it is the relative weakness of the competition in this market which justifies public intervention in the shape of an universal service policy. Competition in the telecommunications sector is therefore accompanied by a policy of 'safeguard'. Clearly, this sector is of considerable importance: it includes essential services and is the bearer of those political expectations symbolised by the term 'information society'. But it is also a market with a very particular structure: it is naturally segmented,[2] and its technology tends to favour a concentration of companies. This renders it hard for competition to bear fruit. The situation of certain clients, the poor and those living in thinly populated zones, even risks changing for the worse as a result of such competition. One may then reasonably doubt whether competition, if left to itself, would be able to meet the three goals given above. This is an argument which can equally be applied to other sectors such as electricity, transport or the post.

Finally, (c) as touched upon above, it is precisely the competitive context which induces the political particularities of the universal service. Thus public intervention to impose a non-profitable service on an operator must be carried out in such a way as to see to it that, in fine, all the operators find themselves on the same footing (cf. point 4).

We may also note that the monopolistic context maintains the same logical links with the concept of functional and organic public service as those raised here between competition and universal service. Firstly, the monopoly context participates in attaining the three objectives given above which are also, globally speaking, those of a functional public service. In particular, other operators cannot enter the market and skim off the cream, which would perhaps imperil the financial viability of the public service operator and prevent him from correctly fulfilling his missions. Further, the weaknesses of the monopolistic context — that, for example, it generates high prices — legitimates public intervention: the regulation of that monopoly. Finally, the monopoly context induces a particular method which is based on the ideas of organic public service and regulation.

2.2. A method and its logic

Universal service policy includes different stages which we shall list under point 4. At this stage we would like to identify the general logic of this policy, regardless of the sector to which it is applied.

Starting from the three logical links covered under point 2.1, we may deduce the essential characteristics that the competitive environment impresses upon universal service policy.

This should first be seen as (a) a monitoring of competition and (b) a correction of its effects. Two examples: firstly, competition naturally induces a price fall. Universal service policy aims in its pricing attitude at monitoring and correcting this movement. In a sense it acts as a brake on the price evolution of the dominant market operator, while at the same time proposing a series of reductions (for the handicapped, for schools, etc...) which otherwise would not have been provided spontaneously. In another sense, competition naturally favours market penetration. Universal service policy monitors this phenomenon by seeing to it that one or more operators are obliged to serve those geographically non-profitable customers.

A third essential characteristic of universal service policy within the context of healthy competition is (c) respect of the same, in particular with regard to the principles of non-discrimination and transparency. Let us suppose that an operator is designated to supply financially non-viable services. Both the method used for this designation — legal obligation, public market — and the operator's remunerative clauses must respect these principles. In this respect we shall observe under point 4, that it is the principle of non-discrimination which lies at the basis of the notion of a 'net cost', central to the idea of universal service, and which guides its application. Funds will also be mobilised to reimburse the operator according to the same principles. In particular, the amounts will be clearly defined and proportional to the contributive capacity of the agents called upon to finance these universal service activities.

2.3. The financing of the European universal service

It is necessary here to clarify a characteristic of European policy which will enable us, at point 3 to distinguish the 'conceptual' universal service from the 'European' one.

In the telecommunications sector, Europe permits the financing of the universal service, for a defined maximum list of services, (European Commission, 1998c: chapter II – point 4.1) by means of funds mobilised from within the sector itself, in accordance with clauses drawn up by the member States. One may observe a historical reasoning behind this: during the era of monopolies the sector itself financed its non-viable activities. The limited character of the list of these services may be justified by the desire to maintain a certain harmony between States with regard to this 'taxation' of the sector for the benefit of universal service.

Despite the limited nature of this list, the subsidiarity principle can be applied more fully than one might think. Nothing in effect prohibits member states from engaging in a more voluntarist policy and envisaging, in parallel, another scope, as long as its financing comes from outside the sector.

'The member States may, within the respect of current Community legislation, submit the provision of telecommunications services to certain supplementary demands. These same should not be financed by means of obligatory contribution from the market operators'. (European Commission, 1998c: article 4).

A member state may, for example, enact a policy of making computers available to schools, hospitals, or local authorities, and finance this from its own budget or that of local authorities, that is to say, by allocating the burden of taxation across the whole economy rather than solely on the telecommunications sector (see our discussion under point 4.1).

3. Definitions of The Universal Service Concept

Our aim here is to present different definitions, in relationship to one another, such as the functional universal service given earlier. We offer two supplementary definitions. The first envisages means as well as finalities, and can be applied to several economic sectors. The second, which applies to the telecommunications sector, presents the definition as understood by the European Union.

3.1. The conceptual universal service

At the 'conceptual' level we intend, following the arguments developed above, to treat as universal service policy:

any policy

- targeting within a list of services the objectives of quality, accessibility and affordability
- within a competitive context and
- according to a method
- which monitors competition
- correcting its effects, while
- respecting the principles of healthy competition, in particular those of non-discrimination and transparency.

Characteristic (1) is the one evoked under point 1.2 of the functional universal service. Note that other aims of a public interest may be envisaged by the legislator, such as protection of the environment, cultural development... The functional universal service only envisages one particular form of public interest (cf. chart 2 infra). It is because the objectives of the universal service are limited that the path of regulated competition may be chosen. Other aims may sometimes necessitate a stronger form of public service 'producer'.

The other two characteristics clarify the context and method intimately linked to the conceptual universal service. In the era of monopolies, the context was that of the monopoly and the method was founded on the idea of an organic public service and its tools of regulation.

Our approach allows a fine-tuning of the distinction between universal service and public service. At a functional level, as we have seen, the concepts intersect. On the other hand, the organic public service is one element of the method intended to achieve the three stated objectives. Furthermore it is compatible with a monopolistic (where the operator needs to be regulated) as well as a competitive context where it is appropriate to apply the method of the conceptual universal service (cf. characteristic (3) supra).

If the political choice of liberalisation is raised, the conceptual universal service may be applied to different sectors such as water, electricity and telecommunications. We may note that the idea of 'Internet in schools' emerges from universal service policy in the ideal sense, even if it lies outside the European scope. This, according to the Directive of 26 February 1998, is not financeable by the sector itself.

3.2. The European universal service

If nothing limits the scope of the universal service in the telecommunications sector under European law, only a part of can be financed from within the sector itself (cf. points 2.3 and 4.1).

Within the context of the European texts, universal telecommunications service policy is to be understood as a policy bearing the three characteristics regrouped supra as well as another, namely (4) whose financing may be sourced from within the sector itself.

Although it is called upon to evolve, the maximum scope of the European universal service is, as we have seen, currently subject to limits: it essentially covers public and private voice telephony, even if a certain subsidiarity is accepted with regard to the intended objectives.

Figure 2 illustrates the relationship between the different concepts evoked. The numbers refer to the characteristics mentioned above.

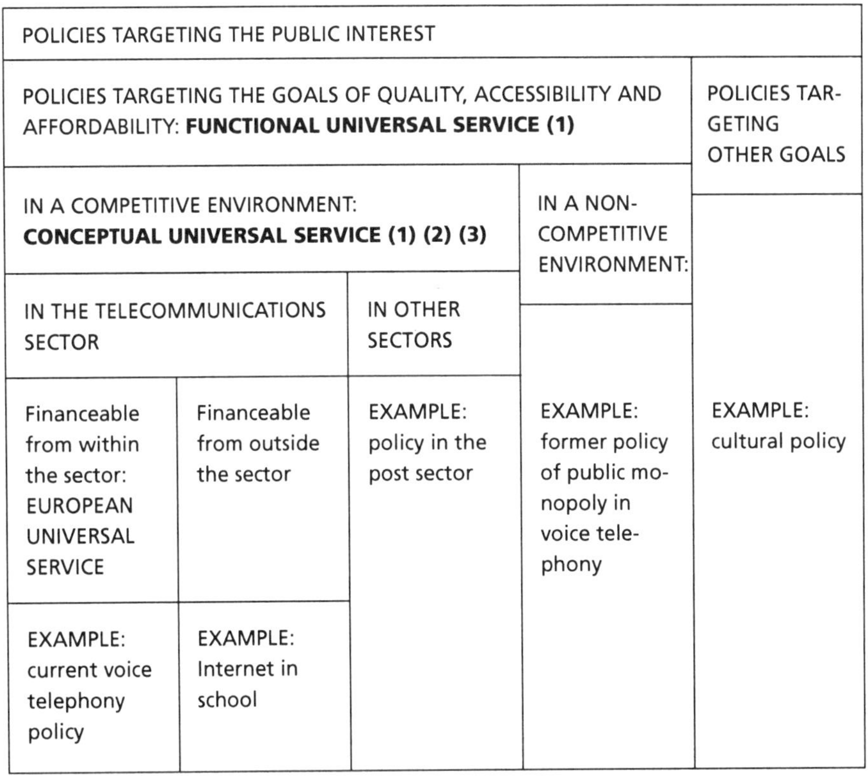

Figure 2 — Universal service concepts

Various clarifications of European universal service policy could be suggested here. It may be viewed as the installation of a minimum platform — both the public interest goals and the list of services involved are limited — suggested to member States, each State being then free to choose to apply a more voluntarist policy.

This European policy may also be seen as the desire to cause the universal telecommunications service to be financed, as in former times, by its own sector, but adapting the method to bring it into line with a competitive environment. In particular, all competing operators are called upon to finance it, according to precise rules.

4. A Space Of Interdependent Political Choices

As we have seen, the idea of the conceptual universal service is intrinsically linked to a context and a method. We should like to examine precisely the ways in which the nation may organise the universal service, assuming the political choice is that of opening up to competition. We should bear in mind that this is but one political possibility among others.

Our objective in this section is therefore to list the various choices to be taken by the legislator in the different stages of universal service policy which we shall identify. We shall see that the fields of choice are in genuine interdependence. The reader will be able to verify that, no matter which sector is under consideration, a universal service policy articulated around the various stages presented here is of necessity. We shall illustrate our argument with reference to the telecommunications sector. We shall discover a space, which is strictly speaking neither legal nor economic and the alternatives of which will be briefly summarised.

The subsidiarity principle can be applied to a certain degree at every stage. The restrictive clauses must, however, respect the logic of the universal service policy as described above — for example, the principle of non-discrimination.

4.1. Determining the scope of the universal service

As the sun rises on the information society, a genuine debate must be held now and in the future, permitting the nation to define the necessarily evolving scope of the universal service concept in accordance with the changes in cultural and social policy. The European Union has recognised this evolutive character elsewhere (European Commission, 1996g). It is therefore desirable to undertake an analysis of social needs and to draw up consultative mechanisms.

If the European universal service scope is limited, the nation is free to envisage, in parallel, an action of greater scope, as long as its financing is gathered from sources outside the telecommunications sector. Thus the European texts cannot be used as an alibi for limited public intervention in this or that member state. We may note that in the USA (often cited as an example concerning universal service) financing is organised from outside the sector. Nothing, therefore, need prevent a member State from drawing its inspiration from this experience or even from going further (Poullet, 1996).

The national legislator must therefore draw up the objectives to be achieved by the different services concerned. In other words, fill out the logical framework presented above under point 1. This is done according to the deficiencies of the market — observed or anticipated (cf. relationship (1) of chart 4 infra). As regards the scope of the European universal service, the texts target public and private voice telephony (including emergency and helpline services) and the telephone directory (European Commission, 1998c: chapter II).

When determining the scope of service domains, society is confronted with its responsibilities: by means of arbitration, goals must be balanced and priorities defined. This is clear in those cases of financing from sources external to the sector: national, regional and local budgets are limited and a universal service policy finds itself in competition with other needs and policies. This arbitration should also precede cases where financing is from within the sector. Basically — although such financing appears less onerous for society as a whole — this is, in terms of tangible realities, merely an appearance. If the sector is solicited by means of some sort of 'tax' as a source for financing the universal service, an eviction effect may be observed: less is produced and less costs more, 'tax' included. An overly zealous extension of the scope of the European universal service may then, in fine, put its very own goals of quality, accessibility and affordability into jeopardy (cf. relationship (9) of figure 4 infra). Allowing universal service policy to be dictated by the effect of a single trend is therefore clearly to be avoided.

Once the scope is determined, the different objectives will be pursued by a series of policies tailored to each other. We suggest, in figure 3, a dissection of these policies to permit us to order them according to their goals. Finance policy, absent here, may be found in figure 4 infra. Its aim is the financing of all these policies. This will enable us to better distinguish infra between global and specific affordability. Our dissection will also give an explicit place to the policy of operator designation and already announce its link with remuneration policy. This will be explained further on.

OBJECTIVES	POLICIES
Quality	Quality oriented policies
Global affordability	Price policy
Accessibility	Policy of designation
	Policy of renumeration
Specific affordability (preferential tariffs)	Policy of designation
	Policy of renumeration

Figure 3: Goals of the universal service and their policy

Note that certain transversal questions arise with regard to these different policies. Take, for example, the issue of the respective fields of jurisdiction of the legislator and the national authority for regulation, or that of the choice of the best adapted legal instrument (law, licence...).

We suggest in chart 4, a unified presentation, situating the goals and the particular stages in drawing up universal service policy. Determining the scope, seen above, constitutes precisely one of these stages. The chart takes the interdependencies into account and identifies — in bold — the elements of choice. This presentation can be applied to any sector, as any political discussion may be seen as relevant to one or other of these stages. Each of these will be briefly presented in its relationship to the telecommunications sector in the continuation of this section.

4.2. Quality oriented policy

When monitoring competition, the legislator may impose service quality on certain operators (cf relationship (2) of Figure 4). This represents technical characteristics as much as adjuncts to the service (helplines, emergency call routing, payment default management and disconnection prevention...).

Several issues are raised here. The majority may be linked to a central question: examining just *how* to impose this regime of quality without discriminating against one or more operators.

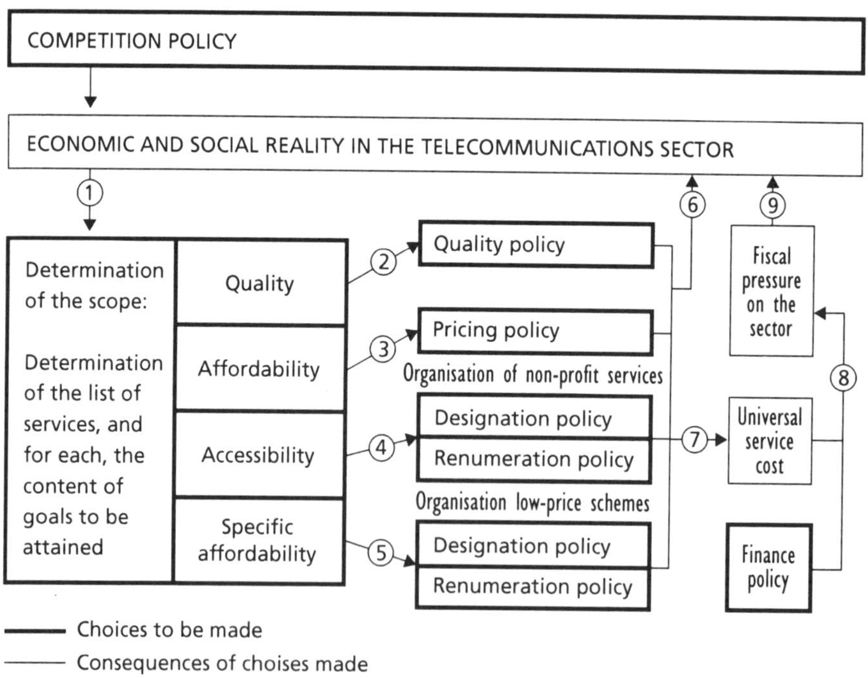

Figure 4:
The development stages of universal service policy and their interdependence

4.3. Price policy

'Global' affordability, that is to say the affordable character of the 'full' price, is one of the expected effects of an opening up to competition. The legislator may, in parallel, pursue a definite price policy[3] (cf. relationship (3) of chart 4 supra) This policy targets price levels as much as price evolution and unicity. Unicity may be seen as a guarantee for price differentiation on the basis of 'supply' issues such as cost differences in the provision of certain services to certain customers in particular geographical regions. This is opposed by those 'demand' issues for special customer groups under different headings (income levels, handicaps, school statute, etc...) which may or indeed should, according to one interpretation of 'equality', be treated in a different manner. This concerns specific affordability, dealt with under point 4.5.

We also note that gratuity, free of charge, is just one of the particular restrictive clauses of price policy, which typically concerns directories.

In telephony, price levels may be regulated for the dominant operator, if the market is still too oligopolistic, by a clause that sets the ceiling or 'price cap' and that may be included in the licence. Unicity, though relative, could also be imposed on every operator for their own prices.

We note that price differences from one operator to another will probably persist: this is due to the freedom of pricing enjoyed in a competitive context. But this competition will also tend to reduce the gaps and to force lower prices. 'Perfect' competition generates a single price. Here, however, the market is geographically segmented and has an oligopolistic structure.

4.4. The organisation of financially non-viable services

Given the currently required quality and the expected price, certain services are financially non-viable for operators. This may endanger the goal of accessibility. We mention the example of customers for certain geographically badly situated telephone boxes; this may also be the case, given the gratuity that is imposed, with the printing and distribution of the private subscriber directories to all consumers.

To meet the goal of accessibility, the regulator must therefore organise non-viable services (*cf.* relationship (4) of chart 4). Two questions must be asked: firstly, concerning the type of designation for operators who provide such services and secondly, concerning their remuneration. The answers to both are, as we shall see, necessarily interconnected.

4.4.1. Designation policy

The regulator may either (1) impose[4] the provision of these services on one or more competing universal service operators, or may (2) for defined services, put all the operators in competition with one another through a tendering procedure: i.e. the regulator will choose the operator on the basis of the sum demanded — the lowest — for the services to be delivered. (European Commission, 1996g)

It should be noted, that this second method, which puts all the operators on the same footing, remunerates them up to the level of a sum revealed by the procedure itself. Remuneration policy and designation policy are thus intimately linked.

4.4.2. Remuneration policy

If the regulator (1) imposes universal service activities on some, an asymmetry of treatment is created and must be corrected. The regulator must therefore deduce, by means of a rigorous system of calculation, the sum corresponding to the net financial charge — net induced receipts — of these services. This net charge, called 'avoidable net cost', is crucial in the European texts (European Parliament, 1997: annex 3) whereby 'avoidable' means 'which would not have been taken on had there been no obligation'. If a remuneration matches this amount, it places in fine all operators, with or without universal service obligations, on the same footing, in accordance with what is required by the principle of non-discrimination.

One can easily see that the drawing up of such a methodology is difficult. It is a matter of establishing the avoidable net cost as closely as possible and without overestimation by precise accounting rules imposed on an operator.

On the other hand, the procedure is less tedious if the regulator has organised (2) a tender for the services concerned: the amount to be remunerated is revealed by the procedure itself. Such a procedure is justified as soon as the situation allows, that is to say, as soon as the state of competition is effective (many market actors being in competition, the sum bid naturally approaches the avoidable net cost) and if the services open to such a tender can be clearly defined and marked off.

4.5. The organisation of preferential tariffs

Once global affordability has been achieved through pricing policy, the legislator may introduce reductions to the full price in favour of certain categories of users (cf. relationship (5) of figure 4. This policy aims at what we call specific affordability, when society as a whole has envisaged, in its conceptual universal service scope, reductions in voice telephony rates for users in very low income brackets, for the handicapped, for schools and local authorities, or a preferential tariff for Internet use in schools.

These reductions, whether or not they fall within the defined European scope, also raise the twin issues of operator designation and remuneration. If they lie outside the European domain, as is the case with the last three examples mentioned above, they must be financed from outside the telecommunications sector. Questions of designation and remuneration will, however, arise in the same way. Similarly, these provisions being by nature different from those targeting accessibility, such as services to geographically non-viable customers, different responses will need to be found to them. In this respect the European texts seem to be unsatisfactory, inasmuch as they ignore the actual logic of the conceptual universal service. This issue needs to be paid particular attention to, but that is outside the scope of this current contribution.

4.5.1. Designation policy

The question arises who should introduce the reductions decided upon by the legislator, some or all operators? The task is to choose the dominant solution according to the principle of non-discrimination between operators and the desire of the legislator to use tarification to favour certain users such as the handicapped and schools...

4.5.2. Remuneration policy

The question is to know if, given the nature of this provision (a reduction), remuneration based on the avoidable net cost is what approaches the legislator's goals best. If this is not the case, a more adequate concept must be suggested: the difference between the full price and the reduction.

4.6. Universal service financing policy

The different policies presented under points 4.2 to 4.5 influence the reality of the sector (cf. relationship (6) of chart 4 supra).

For the expected price and quality, the provisions of the universal service generate in another sense a global sum to be remunerated, the 'universal service cost' (cf. relationship (7) of chart 4 and points 4.4.2 and 4.5.2 supra).

A sum of comparable size must be earmarked via a finance policy, unless this sum is deemed too small under the principle of proportionality. The legislator may consider a priori, for example after consulting the operator responsible for the universal service, that the universal service cost is negligible or non-existant (notably in cases where the non-financial gains, such as those of image, are relatively high). In such a hypothesis, one might delay the introduction of finance policy for part of the universal service, in accordance with the principle of proportionality. This is the case in Belgium where the funds for the universal service will only be released in the year 2000 at the earliest. This, however, does not exempt the legislator from taking care of the other stages of universal service policy. Without there being an anticipatory effort, it seems indeed likely that the finance policy will suffer an induced delay. In particular, the legislator would do well to make a simplified calculation of universal service cost to test the calculation method in order to foresee the level of taxation and to determine the moment at which the universal service cost would become significant. As regards the European universal service scope, the telecommunications sector will be solicited to cover the cost (cf. relationship 8 of figure 4 plus figure 5).

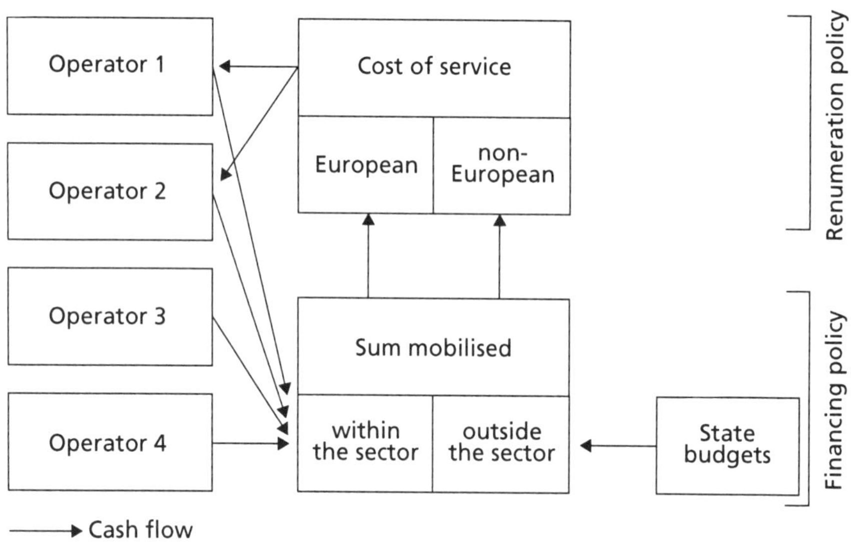

Figure 5: The financing of universal service activities

According to grid 5, special care should be taken that the sum within the sector is raised at the exact time the universal service cost is revealed. A certain rigour should therefore prevail. In particular, the correct provisions must be established with regard to the sum to be raised, the tax base it is to be raised from and therefore the level of 'taxation'. At the same time, the system must show a certain flexibility: it must foresee a safety margin incumbent upon the state budget in the event that the sum raised should prove to be insufficient. Basically, for reasons of legal security, it is difficult to allow contributors to assume ex-post the essential incertitude of submitting to a rate readjustment.

4.6.1. The choice of a financing method

The European Union admits two means of financing from within the sector (European Parliament, 1997: article 5.3). These are the universal service's own funds (fed by the contributive capacity of the operators) and the additional supplement of interconnection charges (a rise in the price of charges levied on operators connecting to the traditional operator's network, which in turn is responsible for supplying the provisions of the universal service). The member States must choose one or the other, or a mix of both of these solutions, in accordance with the subsidiarity principle.

4.6.2. Practical organisation

Whatever system is chosen, the legislator must organise it according to the principle of subsidiarity. In particular, an adequate and precise levying basis must be decided upon for both cases (who has to contribute, what is their contributory base?), as well as the level of participation. In any case, these choices must respect the principles of non-discrimination, legal security and transparency.

4.7. The Interdependence Of Choice

The different stages of universal service policy as sketched here each merit a reflection of their own. On the other hand, as a glance at figure 4 will convince us of, they are clearly interdependent. Let us illustrate this idea. Suppose that the legislator intends to put a more restrictive quality policy into practice (cf. relationship (2) of chart 4). This would involve a rise in the price of services provided and some would no longer be viable. At the given accessibility level, guaranteed by the organisation of non-viable services (cf. relationship (4)), a rise in the universal service cost would be observed (cf. relationship (7)). Depending on the finance policy, additional resources would have to be mobilised: the level of taxation would have to be revised upwardly, thereby stressing fiscal pressure on the sector (cf. relationship (8)). The sector would react (cf. relationship (9)) and 'after tax' prices would rise.

Conclusions

The universal service idea is certainly new, and its application in the telecommunications sector gives rise to questions. These concern the political choices which have caused the emergence of the idea, as much as the parameters to be applied to a concept that is still in the process of sectorial, geographical and temporal development.

This contribution intends to draw up a first assessment of the state of things both on a conceptual and policy level. Too often, the universal service has been understood in terms of its aims; its affordability, its quality and access. Such an 'unanimistic' approach fails to illuminate the underlying political choices: that of a particular method put into practice for the attainment of these goals. This conceptual effort shows that the universal service should be relocated within the domain of public interest. The distinction between the ends served by universal service policy and the political means used to attain them, shows that 'regulated competition' — the solution chosen in the telecommunications sector — is one way among several of which the relevance needs to be evaluated in the light of the sector's inherent characteristics.

The universal service in its currently accepted form, is therefore linked to competition. In the telecommunications sector, as in others, competition may be viewed both as a means to the attainment of the universal service's stated goals and a constraint on the policies envisaged to complement it. In this 'safeguard' policy, it is nonetheless preferred that the state delegates rather than performs the function itself.

The first conclusions of these reflections should not be regarded as definite. In many ways, current concepts (functional public service, common carriage, public utility ...) may well fulfil purposes analogue to those of the universal service. Dissections and regroupings remain to be made.

An analysis of the current conceptual status should not be considered as an effort made 'in camera'. On the contrary, it should be intended to contribute to a better reading of policy which in its turn will considerably enrich the debate. This will apply particularly when the conditions that are linked inherently to a sector and its history, are explained with regard to their influence on the current form of the universal telecommunications service. In the same way, further downstream, an evaluation of the organisational parameters of this policy is intimately connected to the conceptual approach. How is the market that evolves from the chosen option organised, and what are the legal instruments which may improve this organisation?

In the telecommunications sector, opening up to competition is an option and the debate has moved downstream. We may not, however, consider the case closed. Evaluation of the chosen parameters takes place, notably in comparison with the various national approaches and their very concrete effects.

These various experiences should mutually nourish one another if new solutions and legal instruments are to emerge. Similarly, European law should not be considered definite. In fact, it guides some very independent national choices. Reflection on and study of these experiences should show to what degree the European guidelines are either too rigid or too flexible.

When considering other sectors, such as electricity, railways..., the exchange of experiences may also prove a valuable tutor, if used prudently. However, at least two considerations must be respected before effecting such an exchange. The first is that of similarity in service policy goals — this targets the issues of affordability, quality and access. The second is the similarity of the intrinsic characteristics of the sector. Thus, to a certain degree, one may consider that the electricity and telecommunications sectors are both naturally segmented and propitious to the flowering of a concentration of companies. This might justify the adoption of similar policies.

Doubtlessly one may be tempted to apply the universal service solution to sectors such as banking, insurance and education. However, caution should be advised. Such transfers must first clear the hurdle of what we have called earlier the unanimistic vision of the universal service. Of course, it is easy enough to reach an agreement on the intended aims. Still is it not much harder to agree on the methods to be deployed in order to achieve them? Do we for instance envisage increased competition between schools? This choice, which has nothing technical to it, cannot be viewed as the natural result of pronouncing the words 'universal service'.

A certain euphoria for the universal service idea should not blind us to its limitations. It envisages a particular form of public intervention for the attainment of certain goals of general interest. As soon as one begins to speak of the 'information society', policy goals begin to move beyond the characteristics of affordability, access and quality. They reflect a vision that is more cultural, more political and where social cohesion occupies a central position. The desire to extend the scope of political action to include information as an end in itself, should not be subject to hasty judgements. The study of national and sectorial experiences should not be considered here as a defining advantage.

The worst would be to consider the opening-up to competition as a natural element in any public intervention intended to safeguard society's interests. Here again, conceptual reflection (distinguishing ends and means and underlining as much the generalities as the particularities of a sector's experience) should aim not to substitute fundamental choices, but rather to illumine them.

Notes

[1] We shall observe under point 3, that the universal service concept, applied of necessity in a competitive context, may also be introduced into a monopolistic environment.

[2] Consumers are linked to 'geographical zones'. The market is naturally subject to price differences arising on interzonal communications.

[3] Read on this subject the communication on the universal service of 13 March 1996, cited in European Commission, 1996g: 'member States must guarantee that appropriate measures are taken (for example, price caps, targeted tariff grills) ...'.

[4] The regulator chooses the operators on the basis of certain criteria such as, for example, their price structure or their 'geographic' aptitude for the provision of these services.

Pascal Verhoest

The Myth of Universal Service: Hermeneutic Considerations and Political Recommendations[1]

The concept of universal service stands high on the political agenda about telecommunication. Along with the liberalisation of the sector, both in academic and political arenas, the precise significance of the concept was repeatedly discussed. This meaning varies between an 'economising' and a 'socialising' mode. In the economising mode, universal service is discussed in terms of efficiency and distribution of *economic welfare*. The concept then generally consists of three basic criteria, namely the accessibility, the affordability, and the quality of service for all customers. This approach is the one prevailing in contemporary policy documents in the industrialised world. In the socialising mode, the concept starts from a notion of *general welfare* and citizen's rights to access communication resources. In the latter case, universal service is a concept that goes beyond market relations. This 'broad' approach is the one preferred by many progressive politicians and activists. It can be found in claims to elaborate the concept by defining 'additional' service criteria such as the preferential treatment of certain users and the free access of citizens to certain electronic information services.

This article traces the origins of the universal service concept in a European context. It shows that many of our received ideas about universal service consist of historical mystification and investigates how contemporary political players use and reinforce these myths. In that light, the 'broad' conception of universal service, as defended by many progressives, is questioned. It is argued that conceptual clarity could be improved by distinguishing 'basic' universal service obligations (market oriented criteria) from social service obligations (the elaboration of the service on a non-market basis). Furthermore, it is argued that political clarity would be improved by using this distinction. As is documented, the dual meaning of the concept gives it a social connotation that can easily be misappropriated to legitimise a policy which is primarily guided by narrowly defined market economic principles and reduces the possibilities for the provision of services on a non-market basis.

The question about *what* kind of services should be offered on a non-market basis, in turn, relates to the practical, but essential question about who should finance such services: the 'market' or the 'state'. This article shows how a clearer distinction between universal and social service can contribute to the clarification of this issue. It is argued that the provision of (basic) universal service in contemporary telecommunication markets is rather unproblematic. As telecommunication markets mature, and the cost of offering universal service decreases, granting financial compensations for universal provision to (incumbent) operators may be considered superfluous. Rather, it should be investigated to what extent operators can be legitimately called upon to finance meaningful social services. This, in turn, raises the question to what extent operators can be granted public responsibility.

1. The political economy of universal service

The term universal service is closely connected to the development of telephone networks in the US at the beginning of the twentieth century. The term finds its origins in the strategy of the Bell Telephone Company (the forerunner of AT&T) from 1885 onwards, to forestall competitors by pursuing a policy of 'system integration'. This strategy consisted on initially interconnecting competing local exchanges and then absorbing them in the Bell 'grand system'. As Mueller (1994) convincingly argues, AT&T did not aim to extend the service to all. Universal service simply meant nation-wide interconnection. Thus, 'universal' implied everywhere, rather than everyone (Dordick, 1990). The connotation of universal as 'a telephone for everyone' was only added for political reasons: the AT&T credo One System, One Policy, Universal Service served to justify a national telephone monopoly to the Congress.

In Europe, the term universal service has only recently come into use. Its appearance in political parlance coincides with the introduction of competition in network operations. As Garnham (OECD, 1991) points out there was no urge to elaborate explicit service obligations under the former public monopoly, since the state, by its nature, could be supposed to act in the public interest. It may therefore seem surprising that the origins of the notion of universal service can be traced back to the nineteenth century state monopolies in different European countries. As is shown below, this is explained by the fact that the service was not only defined from the perspective of the nature of the role of the state, but with a clear reference to the market.

1.1 A brief history of universal service

In European countries where economic liberalism prevailed in the nineteenth century, like for example in Belgium and the United Kingdom, the government at first turned to private entrepreneurs to operate the telegraph. In the United Kingdom, the Telegraph Act of 1863 established a number of rules for the operators with regard to the affordability and the accessibility of the service. Despite this agreement, the regulation of private telegraph enterprises remained problematic and economic pressures to nationalise the industry increased (Garnham, Joosten & Owen, 1994). The public telegraph monopoly was established in 1869, after market failure had been observed and it was felt that one operator would be more efficient than two or more. Because a private monopoly was judged intolerable, the government was to be preferred to take care of the business (Foreman-Peck & Millward, 1994).

In the United Kingdom, the authorities accepted the idea of nationalisation only reluctantly. The Belgian government adopted a more voluntary attitude than the British and already established the national monopoly in 1850. To a large degree, this is explained by the Belgian inheritance of Saint-Simonism, according to which (economic) state intervention was justified whenever the 'public interest' required it. However, because of the dominating pro-market ideology, the existence of the public monopoly was never fully taken for granted. This was expressed by a careful monitoring of the operator by the dominant class of capitalists and through continuous political pressures to expand the service geographically, to reduce tariffs and, to a lesser extent, to improve the technical quality of the service. In other words, the public telegraph service had to be constructed according to criteria that largely coincide with those of the contemporary notion of universal service (Verhoest & Vercruysse 1994).

Although the above criteria of universality were introduced in Belgium only after the nationalisation of the telegraph, they were set with reference to market economics. According to the prevailing ideology, public enterprise was only legitimated by market failure and the state had to compensate for this by performing better than a private enterprise would under the existing market conditions. Thus, the establishment of telecommunication monopoly was an exemption to the rules of competition, but it was not a derogation of the principles of market economics. The monopoly formula did indeed facilitate the pursuit of universality because it allowed cost averaging and cross-subsidies. However, the emergence of state monopolies was not the result of some a priori commitment to the merits of public enterprise. Rather, it was the result of a relatively careful assessment of the role of the State within the framework of market economics (Verhoest, 1996b). Against this background, it can be counterfactually argued that universality was basically a market-related concept.

It should be noted that, during the nineteenth century, only a small minority of the population could afford to send a telegraph, let alone have a telephone. Despite the occasional rhetoric, the advent to power of the socialist movement after World War I contributed relatively little to the democratisation of telecommunication. The gradual expansion of the service remains closely tied to the constraints of the network economics and to the evolution of general economic conditions. For example, both in the 1920s and the 1950s, subsequent tariff reductions contributed to connect new users to the network. However, these reductions did not primarily aim to democratise the telephone, but to attract new revenue. These extra incomes were required to finance the introduction of new switching techniques in the national grid. This means that, contrary to what is generally accepted, revenues of small business and residential users also served to support the needs of large professional users. In general, pricing would mostly reflect improvements in the cost structure of the operator, rather than the will to give the masses access to the services. In practice, this meant that long term average tariffs diminished at a pace generally just below the rate of inflation (Verhoest, Vercruysse & Punie, 1991). Thus, tariff reduction, geographic expansion and quality improvement, were primarily the result of the increase of general revenue per capita, rather than of a deliberate social policy.

Just as in the United States (Mueller, 1994), the idea of a telephone for every home only appeared in the 1960s and can be considered as an achievement of the emerging welfare state. Preferential tariffs were introduced for certain categories of users and certain vital services, such as emergency numbers, had to be guaranteed for all. In Belgium, for example, preferential tariffs for the less favoured elderly were introduced in 1975. It should be noted that these obligations hardly affected the revenue of the operators. Nevertheless, the European concept of public service was enriched with a social connotation, which transgresses the contemporary notion of universal service.

As Hills (1993) notes, the notion of universal service is a social construct and its definition takes place in response to societal evolutions, and valuable proposals to adapt it to contemporary needs have been formulated (e.g. Hudson, 1994). Because the notion of universal service evolves with society, it is not surprising that the advent of the so-called information society has stimulated a renewed debate on the concept. This can be found, for example, in the claim that telephonic interconnectedness by itself cannot deliver the promise of the information society. In pursuit of a democratic society, Schement (1995) argues, we might ultimately ask ourselves what rights to information belong to all citizens. For example, if democratic participation is to be encouraged in the new environment, citizens must have access to information on the Internet. In this respect, Schement argues that universal service is so important to the information society that it might be better understood as an 'information bill of rights'.

1.2 Liberalisation and universal service

We might also ask ourselves if universal service is the most appropriate concept to indicate this kind of service provision. To use Garnham's (1997) words, universal service has a mythical history. This raises questions about the future use of the universal service concept. It is therefore useful to investigate how universal service is defined in official political practice and how it is being used. The European Commission, for example, defines it as follows: 'universal service means a defined set of services of specified quality which is available to all users independently of their geographic location and, in the light of specific national conditions, at an affordable price' (European Parliament, 1997). Like most contemporary official descriptions of universal service, this definition is based on three pillars: affordability, accessibility, and quality of service. The parallels with the criteria proposed by the nineteenth-century capitalists are striking. However, what do these criteria mean in the contemporary competitive environment? In this respect, another striking parallel emerges.

Indeed, the service criteria of accessibility, affordability and quality of service describe an output that roughly coincides with the classical definition of economic efficiency. Products or services would, firstly, be offered at the lowest possible cost, i.e. reflect optimal production costs (cf. affordability); they would, secondly, be optimally distributed among the members of society given their disposable income (cf. accessibility); and, thirdly, there would be an optimal level of innovation (cf. quality). In other words, in a perfect economy with perfectly 'mature' telecommunication markets, there would most probably be no need for universal service regulation. From this perspective, universal service obligations are only regulatory incentives to favour the effects that a mature competitive telecommunication market can be supposed to have.

One of the elements open for dispute in the above approach is the way it conceptualises the relation between affordability and accessibility. By defining affordability in relation to production costs, and accessibility in relation to disposable income of customers, such a definition indeed limits the possibilities of a socially inspired redistribution of resources in telecommunication. A frequently heard critique of liberalisation is that under the monopoly regime, tariffs were set along the lines of 'value-oriented pricing'. Prices were set with reference to the 'value of service', that is, relative to demand elasticity. Low elasticity means that a rise in price severely cuts down demand, while high elasticity holds that a rise of price will cut only modestly into the quantity of demand. Residential users, who were perceived to have lower price elasticity, were charged less compared to business users with high price elasticity. In this way, 'value-oriented pricing' included the allocation of financial resources originating from major users to smaller users. The same critics argue that today's pressures in favour of 'cost-oriented pricing' — that is prices set, relative to the real cost of providing a service — tends to put an end to this kind of tariff balancing and thus to the social redistribution of incomes from telecommunication operations via this mechanism.

This assumption can be questioned, not only because, as was shown, this mechanism works in two ways. The theory of Ramsey pricing holds that if a service, say residential telephony, has a large price elasticity, then a rise in the price of that service will add little to the firm's revenue. But if the demand for good Y, say professional telephony, is inelastic, then a rise of its price will cut only modestly into the quantity of its demand and so will add a comparatively great amount toward eliminating the shortfall in the firm's revenue (Baumol & Sydac, 1994). In other words, in the long-term, it can be assumed that even in a telecommunication market without universal service regulation, a form of 'redistribution' will continue to exist. It would therefore be erroneous to grant too great social significance to universal service regulation. It may indeed contribute to temper the socially unwanted effects of rate rebalancing. However, just as in the past, it would be exaggerated to interpret this as a deliberately pro-active social policy.

1.3 The politics of universal service

Because of its dual economic and social meaning, the concept of universal service merits some clarification, be it only from a heuristic point of view. Furthermore, political reasons render such an elucidation appropriate. In a Communication on services of public interest (European Commission, 1996h), the European Commission vaunts itself by stating that, in the past, most member states did not guarantee an affordable telephone service for everyone, but this is now being introduced by the European legislation on universal service. Despite the rhetoric, the merits of the Commission are limited. Firstly, as was argued, there was no urge to formulate such explicit guarantees in the past. Secondly, the seemingly voluntary and social attitude of the Commission masks that it has defined universal service in a rather restrictive way. The Commission does allow the preferential treatment by the member states of certain categories of individual users, such as the disabled. It also allows for the creation of a universal service fund to cover up for the cost that an operator may incur by offering universal service. However, the universal service regulation does not apply to any other of additional services that the member states may wish to impose on their operators. For example, when Belgium wanted to introduce preferential tariffs for hospitals and libraries, this was rejected by the Commission with the argument that, under current European universal service regulation, losses incurred by providing such services, cannot be recovered by the universal service fund.[2]

Hence, it may be useful to distinguish social imperatives from universality. Current debates on universal service and the Internet also illustrate this. From a technical point of view, the Internet consists of computers that are connected to the telecommunication infrastructure. All that is required to plug into the Internet is a telephone line with sufficient capacity, as is required and specified under current European universal service regulation. Consequently, from the point of view of 'basic' universal service, Internet is a rather unproblematic network technology. Problems do however emerge when the relation between subscription, computer prices and individual income levels is taken into account. From a social viewpoint, it can then legitimately be argued that the Internet subscription or the purchase of a computer by low-income households should be subsidised. However, such measures clearly do not belong to the realm of classical universal service policies (which, furthermore, generally do not apply to terminal equipment).

2. The economics of service obligations

Studies show that any loss of subscribers as a result of rate rebalancing is likely to be negligible in mature telecommunication markets (OECD, 1991). Rate rebalancing is therefore unlikely to provoke vast amounts of people to drop out of the telephone system. Yet, these studies do not take into account the exclusion that already exists. Indeed, surveys show the existence of social groups with substantially lower telephone penetration and find income to be the single most important influential factor in predicting the absence of a telephone in the home (Schement, 1995). This indicates that there is still a need for universal service measures and, a fortiori, for social service policies. This, in turn, leads to the trivial, but essential question: who, if anybody, should pay for it?

2.1 The limits of competition

Although existing universal service regulation has only a limited scope, proponents of market economics regularly attack it. The magazine Wired, for example, claimed that the 'artificial' imposition of such duties puts unnecessary burdens on economic efficiency and innovation (Browning, 1994). Such popular attacks are firmly rooted in vulgar interpretations of neo-classical economics that assume that only a perfectly 'free' market can guarantee an optimal welfare distribution. However, even supposing that a perfect market would bring about all imaginable social benefits, it is questionable whether such market conditions can be constructed. Indeed, many distortions continue to affect communication markets, particularly at the level of infrastructure.

This is partly explained by historical reasons. The dominance of a single large firm and joint supply of services characterised telecommunication markets. Therefore, the transition towards a more competitive market requires considerable pro-competitive regulatory efforts. In this respect, some have argued that regulation will only be a means of holding the fort until competition arrives (Littlechild, 1983). However, serious doubts can be cast with regard to this assumption. It is not possible to extensively review the literature on this matter within the scope of this article. Instead, a brief reflection of the basic tenets of the concept of 'perfect competition' is elucidated.

According to the definition, three basic conditions have to be fulfilled in order to achieve a perfectly competitive market: firstly, market entry and exit are easy, secondly, there are many small competitors, and, thirdly, firms try to maximise profits. None of these conditions can easily be met in a telecommunication infrastructure market. Firstly, entry is problematic, because the business is characterised by important and significant up-front entry costs. Secondly, because any competitor wishing to duplicate the incumbent's network cannot do so without incurring substantial 'sunk' costs, the number of candidates to enter the field is reduced. Thirdly, as far as the evidence goes, the only condition that is actually met is that of profit maximisation. According to classical market theory, this should have had the effect of improving efficiency, thus bringing down prices. However, profits are also put to effect for more 'perverse' reasons, such as cross-subsidisation and cost averaging techniques that allow large operators to develop predatory pricing strategies that continuously hinder competition.

Economists have asserted that in the absence of effective competition, firms can still be stimulated to produce efficiently if they are *threatened* by potential entrants. In this case, the market can be said to be 'perfectly contestable'. However, just like the concept of perfect competition, the concept of perfect contestability is also a purely theoretical construct. It presupposes the absolute absence of barriers to entry; a condition that is never actually met (Baumol & Sydac, 1994). Telecommunication enterprises are generally well aware of this and will therefore continue to display anti-competitive behaviour. Therefore, the telecommunication market will continue to be dominated by a few major players and regulation has to keep operators in line with fair market practices. Regulators can respond to this, either positively by discriminating in favour certain service providers and/or negatively by discriminating against certain other operators, in order to limit the possibilities for abuse of dominant position. In both cases regulation is a means of creating 'virtual competition', that is, of creating a situation that simulates market conditions.

In sum, the uncompromising advocacy of competition will not necessarily guarantee economic efficiency. Economies of scale and scope, which are most strongly felt in the infrastructure business, reduce the possibilities for economically sound competition for some telecommunications services. Because there are limits to effective competition, in some cases, monopolistic or oligopolistic rivalry may even be preferable to theoretically more 'perfect' forms of competition (Mansell, 1993). Rather than creating the conditions of 'perfect' competition or contestability, the regulation may then be aimed at keeping firms in line with acceptable economic and social standards, be this by the imposition of certain service obligations. In this case, a major problem is to define socially and economically acceptable service obligations. Here too, classical economics are of limited value.

Because the conditions of perfect competition are never actually met, there is no way of knowing what this presupposed ideal output would be. Besides, even if the telecommunication market would be highly competitive, there still would no guarantee that the optimal output would be produced, since perfect competition requires that all markets are perfect. Moreover, even if the ideal of a perfect market could be approximated, the capacity of that market to produce the socially desired outcome can still be questioned. One of the reasons is that the actual expectations of social agents with regard to communication services do not necessarily correlate with some 'objective' long-term economic optimum, but tend to vary substantially according to their immediate social positions and interests. Thus, even if the output competitive markets would be Pareto-optimal — that is, an increase in welfare would make people better off without at the same time making anybody worse off — that output will not necessarily be regarded as equitable (OECD, 1991). From this point of view, the relative distribution of welfare among citizens is more important than the absolute increases in welfare.

2.2 Capitalising on competition

As was argued, the fundamental legitimisation of the former telecommunication monopolies followed from market failure. Telecommunication was considered to be a natural monopoly, that is, production would be more efficient if taken care of by one firm. Therefore, a single operator was more apt to achieve the goal of universal service. In other words, a monopoly position was granted on the condition that certain obligations were fulfilled. From the point of view of the regulators, such service obligations were an instrument to counterbalance the monopoly power of the operator. From the point of view of the operators, these obligations were the price operators that they had to pay for holding a monopoly.

This rationale is now being questioned as a result of liberalisation. There is indeed an apparent contradiction in a policy that, on the one hand, promotes competition among operators by compelling firms to operate under more stringent market conditions and, on the other hand, imposes 'non-market' obligations on certain operators. From a strictly market-economic point of view, such non-market obligations affect competitive behaviour and distort market conditions. However, as was argued, telecommunication infrastructure markets are far from being fully competitive and it is therefore legitimate to continue to impose some obligations on network operators. The effect of liberalisation is thus not that it renders the old rationale obsolete but that it merely complicates the calculation of the way in which the costs of providing the service have to be distributed among several market players.

When subjected to service obligations, incumbent operators generally argue that financial compensation is required to cover for the losses incurred by newcomers, which do not have to fulfil the same obligations. This claim sounds reasonable, since newcomers often cream off revenues by concentrating their business on the most profitable parts of the market. This reduces the possibilities for cross-subsidisation and cost averaging that contributed to universally expanding the service under the monopoly regime. However, there are important counter assertions to this claim. A study by the British Office of Telecommunication (OFTEL) shows that the cost of universal service is not as high as incumbent operators would like to believe.[3] Moreover, there is reasonable evidence to believe that, as telecommunication markets mature, the costs of providing universal service decreases. If the costs of universal service are becoming marginal, it may make sense to ignore them. Besides, there are also immaterial advantages for an operator to be a universal service provider. It has been suggested that providing 'full' (universal) service is an appropriate marketing strategy, because it creates a positive brand image for the concerned operator (Neu & Stumpf, 1997). Furthermore, it may also be politically advantageous to be the (sole) universal service provider because this may inspire some compliance of the regulator.

Despite these findings, governments and regulators have been sensitive to the arguments of the incumbent operators. The European Commission, for example, allows for the creation of universal service funds. If an operator can show that the provision of universal service causes a net revenue loss for one operator, other operators can be called upon to financially compensate for this. However, the effects of this policy are likely to be limited. As was demonstrated, universal service obligations have been defined in a relatively restrictive way. Moreover, the costs of providing 'basic' universal service are likely to be almost negligible in mature economies. The question can therefore be posed whether a part of the revenues of the telecommunication business, which are known to be considerable, could be used to finance additional social and public services?

Such a policy is likely to be rejected by telecommunication operators. Appealing on classic economic theory, they are likely to reason that the provision of such service is, literally, 'none of their business'. Social and public services should be financed with general tax revenues, rather than by imposing unnecessary burdens on firms which already have to cope with fierce international competition. However, classic economic theory itself contradicts such assertions. In this regard, Melody (1997a: 13-27) mentions Adam Smith who, in *The Wealth of Nations,* recognised that many of the 'common callings' constituted a class apart and his laissez-faire proposals could not be expected to protect the public interest.

Put this way, one of the founding fathers of market economic theory provides us with a moral ground to impose service obligations. However, it still does not provide a solid economic argument on the basis of which to decide whether these obligations should be imposed on the state or on the firms. The relevant question in this respect is: should the service be paid for by the customers or with tax revenues. If a service is judged essential for society at large, it may be legitimate to subsidise its provision with tax revenue. However, such a policy is not necessarily equitable because the tax is also imposed on none users. The imposition of service obligations on the industry may be more equitable. As was argued, this creates a two-way stream of redistribution among users. Because this stream is steered by factors of demand elasticity, it is defendable both in an economic and in a social way, provided the necessary regulatory correction and control mechanisms are installed.

2.3 The future of universal service

The remaining question then is normative: what kind of services can or should be considered in contemporary society? From the point of view of classical economics, this question can be addressed by looking at the positive externalities of the telecommunication business. These benefits are difficult to estimate in a conventional economic way because this requires looking at wider societal issues (Graham, Conford & Marvin, 1996). This can be historically illustrated. In the nineteenth century, the telephone was mainly used for professional purposes. As a means of social communication, it was considered to be a luxury good. The striving toward universality for the capitalist class largely sufficed to fulfil the social needs of that period. During the twentieth century, time-space distanciation (Giddens, 1990) in society increased and the telephone became a vital instrument for all social strata. Pleas for the introduction of preferential tariffs correspond to that epoch. At the turn of the twentieth century, the advent of the so-called information society renders people more dependent on information and consequently the matter of access to these services is put on the agenda. In sum, during history the realm of services was elaborated as their social significance accrued. However, as was argued previously, very often there was a significant gap between rhetoric and practice.

This, in turn, raises questions about which measures are appropriate to ensure that political intentions actually materialise. In this respect, a number of policy measures can be proposed. Firstly, 'basic' universal policy regulation should ensure equal access to new services, such as the Internet. Regulators in industrialised countries have generally taken measures to this effect. This is also the least problematic since the remaining technical barriers are likely to erode anyway, as networks are being updated. More stringent are problems of social exclusion. These require, firstly, measures to bring the telephone into the homes of those that can still not afford it and, secondly, measures to ensure that the less wealthy too will have access to essential electronic services. Such issues are generally more problematic in contemporary universal policies. Therefore, a more pro-active policy is needed to ensure that every citizen can access the telecommunication system.

The problematic nature of the universal service concept becomes even more obvious when applied to the field of information provision. At stake on this level is not access to hardware but the availability and affordability of relevant *content*. This concern is alien to universal service policies, which were historically designed only to provide physical access to the network. The preoccupation for public information provision is much more acquainted to mass media theories that are concerned with the provision of information as a 'public good'. These theories have generally considered the universality of signal carriage as a necessary precondition. In this sense, the concern for public information goes a step further than concerns for universality and social justice in infrastructure provision. The convergence between mass media and telecommunication and the advent of a so-called information society may therefore require the application of this public service logic to the telecommunication sector, as part of the social information and communication system as a whole.

Such measures would contribute to improve the availability of content and as such, they are complementary to a policy aimed at improving accessibility. However, such a policy is still only geared towards physical access, which is to connect with the available information and communication resources in society. It does not hold any guarantees that people have the capability of actually mastering these resources. Therefore, it is useful to distinguish physical access from dispositional access, that is the cultural, intellectual and practical dispositions required to function in an information-rich environment. If it appears that these dispositions are unequally distributed in society and prevents people from fully participating in society, policy will have to address these issues too. The latter obviously requires an information and communication policy that not only encompasses sector-specific and social policies, but also such areas as education and culture at large.

3. Conclusion

Because of its different connotations, the notion of universal service can easily be misused for political purposes. Proponents of a narrowly defined (market-oriented) approach to universal service can easily claim to support a humanitarian policy, while in fact only a 'basic' service is promoted. Alternatively, supporters of a humanitarian approach can positively claim to pursue a social policy, while in practice, only relatively small corrections to the market are made. In other words: the broadening of the concept of universal service does not improve political clarity. It renders it more apt for ideological misappropriation. Conceptual precision would contribute to the elimination of this bias. This can be achieved by using different concepts to denote the different levels of significance contained in the 'broad' definition of universal service. When looking at the historical evolution of the concept, two levels can be distinguished. The first level is that of 'basic' universal service and refers to the correction of market imperfections. The second level is that of a 'social service', that is, a preferential treatment of certain categories of users. The conceptual distinction between universal and social services proposed in this article may contribute to a more careful examination of the role and the tasks of the private and public sector in telecommunication.

The aim of this article was not to provide a decisive answer to the questions of which minimal services should be provided, or by whom, although some suggestions to this effect were made. By means of conclusion, some general guidelines for future reflection can be suggested. The question regarding which services should be guaranteed cannot be posed solely in market economical terms, but instead requires discussion of the significance of the industry for society. Neither should the question of which player will finance the service be answered in dual terms, by opposing the private to the public sector. Neo-liberal regulation tends to treat firms as individual private entities. However, as was argued, even within a classical economic reference frame, private enterprise can be held responsible for important public missions. Consequently, there is little contradiction in promoting economically and socially sound forms of competition and at the same time complementing market mechanisms where these do not suffice to fulfil societal expectations.

The argument can even be carried a step further. Given their important role in society, it may be more appropriate to treat communication businesses as public agents with a responsibility toward civil society. This, in turn, requires a breakdown of the traditional political and economic boundaries between private and public sectors and constitutes an ambitious and long-term 'reformist' political agenda in itself. Such an agenda will also have to take into account the way in which sector specific issues interact with social, cultural and education policies. Thus, establishing a democratic telecommunication policy as part of a broader communication and information policy, requires a profound reflection of the role of information and communication in the constitution of society.

Notes

1 This article is based on research carried out in collaboration with the centre for Studies on Media, Information and Telecommunication (SMIT), Free University of Brussels. This research is financed by the science policy unit of the federal Belgian government's Interuniversity Attraction Poles (IUAP), Phase IV, Research and Interdisciplinary Evaluation on the Information Society: Networks, Uses and Roles of the State and is carried out by SMIT (VUB), CRID-CITA (FUNDP) and Lentic (ULg). Jean-Claude Burgelman (SMIT) and Jos Leijten (TNO) are acknowledged for their precious comments on this article.

2 Based on interviews with different representatives of the Belgian regulatory authority, the Belgian Institute for Posts and Telecommunications (BIPT).

3 This study estimates the cost of providing universal service in the United Kingdom to be between £0 and £40 per year, depending on assumptions OFTEL. (1995). *Universal Telecommunications Services: Consultative Document on Universal Service in the UK from 1997.* Lonon: HMSO.

Anders Henten

Regulation in a liberal, international and converging communication market

Introduction

During the second half of the 1990s, the liberalisation of telecommunication markets in Europe has come a long way — especially with the 1st January 1998 deadline. However, a number of market and regulatory issues are still pending. This applies to the development of actual competition in the markets — with operators challenging their incumbents. It also applies to the regulatory questions attached to the emerging convergence of formerly separate technological and economic areas in the communication fields. Furthermore, it applies to the generality/specificity of telecommunication regulation — which could also be dealt with in terms of the degree of regulation. Moreover, there are questions regarding the proactivity vs reactivity in regulation and the national/international dimension of regulation. In the present paper, three of these questions are dealt with:

- The degree of regulation
- National/ international regulation
- Regulation of converging media

1. The degree of regulation

This question can be phrased in two ways: the degree of regulation or the generality/specificity of telecommunication regulation. There is a common understanding that some kind of regulation is needed, the degree and specificity of regulation, however, is a point of contention. There are those who argue that only a very limited regulation is necessary, that a 'normalisation' of telecommunication markets should take place, and that telecommunication should not be subject to any special regulation. With a focus on the long term, this is the prevalent view. However, with regard to the short term, this point of view is, ironically enough, mostly defended by representatives of the incumbent operators that formerly were protected by monopoly rights and who have an interest in removing regulations of a so-called asymmetric character that seeks to provide a level-playing field for new entrants in the market. They have the market power of the well-established operators to take advantage of telecommunication markets with only very limited regulation.

And then there are also those who contend that the conditions of competition should be regulated with the purpose of favouring new entrants in order to create a real competitive situation in the market. This point of view is mostly defended by representatives of the new market entrants (and by a vast majority of academics and political administrators). Ironically enough, this applies not only to the 'really' new entrants establishing telecommunication operators for the first time, but also to the representatives of the old incumbent national monopolies establishing subsidiaries or entering joint ventures in foreign countries. Apparently, opportunism prevails: at home, arguments are put forward in favour of a non-regulated regime. Abroad, asymmetric regulation is defended.

However, there seems to be a growing general opinion that this kind of asymmetric regulation has reached its peak and that a (new) deregulation is needed. This opinion can be based on two assumptions. One, that competition has already advanced to such a degree that asymmetric regulation can gradually be slackened. Two, that there are dangers that a politically guided competition may substitute 'real' competition. The degree of regulation is seen as a curve where regulation must first increase and then decrease.

Apart from the two above-mentioned points of view contending no regulation or provisional regulation in order to create competition, there is a third point of view arguing that a specific telecommunication regulation will 'always' be needed. In contrast to the two former positions, this position is not only based on an evaluation of the competitive situation in the markets but also on other societal considerations such as public access to communication services (universal service) and access of operators to limited resources (frequencies, rights of way and numbers). More broadly speaking, this point of view focuses on telecommunication as a networked social resource where utility is the greatest if it is interconnected and if the largest possible number of citizens is connected. This applies especially in a situation where communications is said to be of increasingly central importance, i.e. in an emerging information society.

This third point of view is — with its emphasis on the central importance of communications in society — related to a fourth point of view that however has another dimension than regulation of just telecommunications. This point of view is related to the emerging convergence between formerly separate communication areas — telecommunication, broadcasting, information technology and entertainment/ publishing. The question of regulation of converging media will be taken up in a later section in this paper. The point here is that new regulatory questions will arise with convergence and that this affects the question of a more or less specific regulation of communications.

2. National/ international regulation

Telecommunications has been international since its inception. Still, for approximately 100 years, telecommunications was organised as national monopolies, and all that was internationally regulated were the interfaces between the national monopolies. This has been catered for by the International Telecommunication Union (and its predecessors) since 1865, regulating primarily the technical interfaces (standards) and international payments (accounting rates).

With the liberalisation of telecommunications it has become necessary in international agreements to go deeper into the national settings and organisation of the markets. This is what has happened primarily through the World Trade Organization with its agreements on telecommunications in 1994 (value-added services) and 1998 (basic services).

In Europe, there is an additional question concerning the possible setting-up of a common EU telecommunication regulatory body – called the European Regulatory Authority (ERA). Today, directives are issued by the Council or the Commission in collaboration with the European Parliament. These directives have then to be transposed into the national laws of the member states, and national regulatory agencies are supposed to supervise their implementation. However, the question has been raised whether this is sufficient and whether a Community-level regulator is necessary.

In accordance with EU ONP-directives from 1997[1], the Commission in its review of the implementation and consequences of these directives has to include an evaluation no later than end 1999 of the value added of setting up an ERA. The question has been on the agenda for some years but clear arguments against or in favour have not really been discussed. Instead there now seems to be a growing common understanding among national regulators and the European Commission that this is not a central issue, either because it is not necessary, not feasible or not desirable. There are indeed serious problems involved in setting up a common European regulator. The matter certainly deserves to be debated explicitly instead of just being put aside on grounds of more or less implicitly understood difficulties.

The basic question must be whether a European Regulatory Authority is necessary — a question that has two aspects: the first is whether there are problems needing international regulation and the second is whether existing institutions are sufficient. The first aspect must be answered affirmatively with some reservations. The fundamental argument is that a common European internal market needs a common European regulator. However, the European telecommunication market is not yet really a common internal market, but a market consisting primarily of national liberalised markets where operators from other countries can extend their tentacles and they are still, first and foremost, national markets. In such a situation, national regulation may be sufficient.

Nevertheless, there are areas already where national regulation only is not enough. This applies, e.g. to intra-European international communications where tariffs are exceeding by far the long-distance interstate tariffs in the US. The accounting rate system is at play here but should not account for all the blame. The fact is that operators are interested in keeping international tariffs high as long as possible. Another area is mobile communications that, because of its nature reaches, international dimensions more quickly than fixed-line communications. An interesting point is, for instance, that cellular operators circulate nationally bound traffic abroad in order to circumvent high national interconnection tariffs — so-called tromboning. Such problems can, in principle, be solved and may even be solved by the existing regulatory structures. The question is whether this is the most efficient way of solving them.

More fundamentally, the question is whether the existing structure with primarily national markets will continue to exist or whether a truly common internal market will develop. If a really common internal market develops, the appropriate answer is not to rely solely on national regulatory bodies.

Maybe, the question should also be formulated in terms of desirability. If one wants to move in the direction of a common internal market, then a European Regulatory Authority may be a useful tool. But if one wants to halt this process, there may be good arguments in favour of not establishing a common European regulatory agency.

The second aspect of the question regarding the necessity of an EU regulatory body is whether there are institutions already that cover such tasks. Many institutions could be mentioned, but only 4 will be briefly dealt with here. One is ECTRA (European Committee for Telecommunications Regulatory Affairs) under CEPT (the European Conference of Post and Telecommunications Administrations) where national regulatory authorities meet to discuss and coordinate regulatory measures. Another is ETO (European Telecommunications Office) that was set up by CEPT with the purpose of providing an administrative framework for the implementation of a one-stop shopping arrangement in relation to licensing and of studying questions in relation to numbering. A third institution is the Independent Regulators' Group (IRG) that has been set up by European national regulatory agencies with the purpose of creating an informal network of European regulators for debates and the exchanges of experience regarding problems in the processes of liberalisation.

The last and by far the most powerful institution is the Commission itself. The Commission is a strange creature in terms of traditional divisions of power among state institutions. It is both an administration of civil servants and a legislative (directives) institution. The Commission (and in this connection DG XIII and DG IV) could, in a way, be considered as a highly proactive regulatory institution basing its intervention on article 90 of the Treaty of the European Union. However, the Commission is more of a hybrid between a public administration and a legislative body than an independent regulatory institution. But no more do the other institutions mentioned fully cover the task of a European regulator – with all due respect for their many virtues.

Consequently, where does that leave a European Regulatory Authority today? The Commission does not seem to provide much impetus to the question at the moment. In an article in the journal InterMedia, Rex Winsbury (1998) speculates about the reasons for the lack of initiative in support of the establishment of a European regulator. He states that 'the logical thing would be to abolish DGXIII and fold it into a new regulatory body'. And he continues to argue that no bureaucratic structure would recommend its own abolition. On the other hand, DGXIII might 'find a long-term safe haven' by transforming itself into an EU regulatory body (Winsbury, 1998: 34). Perhaps, these pros and cons are close to a correct analysis of motives in the Commission. At any rate, it is certain that support for a European regulatory body is not to be found in the Commission but in Parliament.

Whereas the Commission may be reluctant to yield power to a new regulatory body, the Parliament may see a regulatory body as being more in line with traditional constitutional divisions of state power and competencies — in line with the preference of the Parliament for basing directives on article 100 of the Treaty of the European Union instead of using article 90 on which Commission directives in the telecommunication area have been based.

Such differences of opinion between the institutions of the European Union may be characterised as bureaucratic skirmishes and mere fighting for power. Yet, they may also be founded on more legitimate opinions concerning the best ways to develop the institutions of the EU generally and the institutions with influence on communications in particular.

However, the problems in the creation of an EU regulator do not end with differences of opinion between institutions of the Community. The member states are not enthusiastic about a European regulator either. They prefer the application of the principle of subsidiarity in this area. In spite of the intention that directives should result in similar legislation in the different member states, neither the specificities nor the regulatory interpretations are always the same. Nevertheless, it is generally felt that such problems can be solved by co-ordination and exchanges of experience among national regulatory authorities as, for instance, with the Independent Regulators' Group. Moreover, the drawbacks of establishing a common EU regulator are, at the moment, considered as bigger than the problems caused by the differences in implementation and interpretation.

It should also be remembered that most of the national regulatory agencies have just been established and have not yet entirely developed their role in the regulatory processes. They will be reluctant to find themselves already in a new a subordinate role in relation to a European regulator.

Finally, there would (or will) be major difficulties in finding the right constitutional formula for the establishment of an EU regulator. There are already some pan-European agencies of different kinds, but it would be the first pan-European body regulating industry.

All in all, it seems highly unlikely that a review of the ONP directives by the end of 1999 will result in a proposal to establish a European Regulatory Authority. It may be a 'brave' position to argue that a common internal telecommunication market in the EU requires a pan-European regulator. But it is not very realistic to assume that there will be sufficient support for it.

3. Regulation of converging media

The basic idea here is that if different technological and industrial areas converge, so must the political governance of these areas. In the communication field, telecommunications, information technology, broadcasting, and entertainment/ publishing are converging technologically and economically in different ways and only to some extent. Telecommunications have traditionally been subject to a special regulation of accessibility, interconnection and distribution of scarce resources. Information technology has only been subject to basic industrial regulation. Broadcasting has been regulated on grounds of scarce resources (frequencies) and content (public service). Entertainment/publishing has been regulated, primarily, in accordance with the limits of free speech. When these areas tend to converge, the question arises whether and to what extent public regulation of these areas should also converge. Is it, for instance, necessary to look at content questions in relation to telecommunications?

Dwayne Winseck (1998a) argues that political governance does not come second to technological and industrial developments in the communication field. An important reason for the separation between different areas in communications has been a political desire to block cross-media ownership between different communication areas. This is certainly true; there is however no doubt that digital technology has provided a new technological basis for convergence in the communication field. At any rate, the convergence pressure from industry has vastly increased in the past decades, renewing the question of political convergence in the communication field on the political agenda.

There are a number of areas where the question of political convergence is relevant. This applies for instance to:

- Interconnection
- Universal service and access
- Content
- Cross-media ownership

Interconnection may take place at both network, service and application layers. A case in point, at present, is Internet telephony which in some countries (primarily the US) is not considered as a real telephone service and is, consequently, not regulated as traditional telephony. Interconnection at the application layer does not, at the moment, have the same public attention. However, in the coming years it will be increasingly important that media applications are to some extent interoperable.

Universal service is, at present, confined to different telephone network and service facilities. However, with the development of new electronic media the question is raised whether the universal service concept should be extended to new network and service facilities. In the US, there is already an arrangement where telephone companies are obliged to service educational and health institutions at favourable rates with respect to new networks and services.

The reason for this is partly that taxing (providing funds to public institutions among other things) is a very sensitive issue in the US. However, in most countries there is a general feeling that new cleavages between information-rich and information-poor should be avoided, possibly leading to extensions of the universal service obligation to encompass new network and service facilities.

Another aspect of this question relates to access. There is access in the network sense (which is here termed universal service), but there is also the question of access to information — both in the private and the public domain. If citizens are to take part in developments in the upcoming information society, public regulation has to see to it that citizens have the widest possible access to information, first and foremost, in the public domain but also in the private domain.

Content regulation has often dealt with at least two questions: one, regarding prevention of unlawful and abusive information. Another, regarding public service laying down some demands on certain 'public service' broadcasters. When information can be broadcasted not only on traditional one-to-many broadcast media, the question is raised whether such content regulation should also apply to the new hybrid media. In the case of unlawful and abusive information, there is no doubt. The same rules apply to new media. However, should there be public service obligations on some information providers on the new interactive networks?

Cross-media ownership is not a new issue. As emphasised in the above-mentioned book by Dwayne Winseck (1998a), regulations in many countries have for decades prohibited cross-media ownership. Such regulations are presently being removed, but the question is still there — and now more forcefully than ever because of the possibilities created by digitalisation — whether cross-media ownership should be allowed. The reasons put forward for relaxing the regulations on cross-media ownership is that this will help develop the synergies between the converging communication areas. The relaxation is portrayed as a natural consequence of the new technological possibilities. Still, one should not forget that there is a political choice to be made. Specific political priorities do not automatically follow from technological possibilities.

Questions on interconnection, access, content and ownership have to be reconsidered in light of technological and industrial convergence in the communications areas. And there are many others, for instance with regard to consumer protection in the upcoming e-commerce environment, to privacy protection and to new problems in relation to intellectual property rights. There are enough questions to be dealt with in the coming years in relation to new communication media.

A specific aspect of these questions is whether regulatory institutions should also converge. In most countries, regulation of the different communication areas has traditionally been dealt with by separate regulatory or administrative bodies. It is an important question and has also been discussed, for instance, in the UK. There is however no reason to take the institutional aspect of the question for primacy. The most important aspects are related to the content of the matter. Institutional consequences come second.

4. Summary

The purpose of this article has been to show that there are still many regulatory issues to be solved in the communication field. Many issues have already been settled during the past five to ten years. The telecommunication markets have been liberalised in most European countries and competition is advancing. Nevertheless, a number of questions are still pending.

Three of these questions have been dealt with here: the degree of regulation, national/ international regulation, and regulation of converging media. Regarding the degree of regulation, the question is for what reasons and to what extent a specific telecommunication regulation is needed. The section on this subject does not have an explicit answer to the question, but implies that if competition is to be strengthened and if considerations other than creating a competitive market are to be taken into account, a specific communication regulation has to be in place and must be developed.

The second section concerning the establishment of a pan-European regulator analyses the arguments in favour of and against such a regulatory body and it presents the interests and positions of different players with an influence on this development. The conclusion is that it is not likely that a common European regulatory agency will be established soon, even though there are good arguments in favour, especially if the goal is to create a truly common internal telecommunication market in the EU.

The section on convergence argues that a public regulation has to match the convergence developments in the communication field. Digitalisation has provided the technological basis for an increased pressure for industrial convergence and regulatory measures to reflect this. Certainly, the most comprehensive and difficult task in the coming years will be the regulatory issues in relation to converging media.

Notes

1 97/51/EC and 97/33/EC.

Paul Belleflamme

Market Regulation: Synthesis of the Debate

'The Need for Time- and Sector-dependent Regulation'

The issue of market regulation in the information and communications technologies (ICT) sectors encompasses two intertwined questions. First, is there a need for regulation? Secondly, if yes, what kind of regulation is needed? Broadly speaking, the participants to the workshop agreed to answer in the following way: there is indeed a need for regulation but, to be effective, this regulation must be both sector- and time-dependent. Let us expand on this.

In his introduction, Eric Van Heesvelde briefly recalls a series of problems (such as connection, convergence, or unbundling) that prove the existence of market distortions and therefore establish, in a pragmatic way, the scope for market regulation in the ICT sectors. He acknowledges, however, that the issue of the exact form such regulation should take, remains. To address this issue, one needs to take a stand on questions such as: What do dominance and fair competition mean in these sectors? One also needs to assess the coherence of the legal treatment of these questions by the European Union Member States. Eric Van Heesvelde invites the different speakers to shed some light on these intricate issues.

Jos Leyten takes the plunge and presents his own ideas on what direction market regulation should take. He addresses general issues about competition in communications (viewed as means to interact and to convey messages) in a converging world. He starts by identifying three consecutive stages in the development of a network: in a first stage, there is only a single channel and competition is for capacity (this is so for the Internet); in a second stage, there are multiple channels and competition is for messages and messengers (this is so for broadcasting and cable); in a third stage, there is a chain reversal, demand starts to dominate and competition is now for services (this is so for telecommunications). According to Jos Leyten, the main challenge for regulation is to select policies that fit these different steps; in other words, *time-dependent regulation* is called for. In doing so, policies have to be guided by two watchwords: 'unbundle' and 'think networks'. First, unbundling (i.e. the separation of the network and service layers) and the provision of open network technologies are necessary because they allow service providers to build their own systems and as such give rise to a potential increase in disruptive innovations. Second, thinking in terms of networks means the end of traditional hierarchies through a move towards self-organising systems (i.e. clusters like the Silicon Valley); it also means seizing the opportunities raised by interaction between buyers and sellers, or between servers and clients.

After these general principles, the issue of market regulation is tackled within the specific field of electronic commerce. Pascal Verhoest presents some speculations on the future of e-commerce; his research agenda has an interest not only for scientists, but also for public authorities and for commercial agents. He describes first the opportunities of e-commerce: by making information cheaply available, e-commerce has a potential for alleviating the information imbalance between sellers and buyers and to lead to 'more perfect' markets; it may also increase efficiency through reduced co-ordination and transaction costs. The achievement of these opportunities might, however, be hindered by a couple of significant obstacles. The first obstacle concerns the design of the software: most of the e-commerce software applications are proprietary and are designed to enhance the sellers' capacity to gather information; thus, far form reducing the information imbalance between the two sides of the market, e-commerce tends instead to increase it. The second obstacle has to do with the reduction of the personal context in e-commerce: because of the difficulty of maintaining stable buyer-seller relationships, firms have to rely on their reputation to keep consumers; therefore, firm asymmetry is likely to increase in electronic markets. In this context, what strategies should firms follow? A clear objective is the customisation of production; firms can use e-commerce to become more responsive to individual demands; in doing so, they are able to set up more personalised relationships with their consumers and increase their switching costs. The pursuit of such strategy is likely to induce opposite movements within the value chain: retailers have an incentive to integrate vertically upward, while producers have an incentive to by-pass retailers. In any case, there seems to be room for new intermediaries (the so-called electronic traders).

On the basis of this analysis, Robert Queck addresses regulation issues related to electronic commerce. His main question is whether a new legislation is needed to tackle this new transaction mode. In his view, e-commerce raises many old questions, which suggests that existing legal tools would work. Yet, new problems arise which need new responses: there is indeed an increase in the geographical scope of commerce, as well as new dimensions of risks (such as privacy). Moreover, since trust is a pre-requisite for e-commerce, a strong and stable legal system is necessary. Suitable legal responses have thus to be found with respect to contracts (formation, signature, performance, electronic payment), copyright, liability, and tax matters. In some instances, self-regulation might bring faster and more adapted solutions along; but self-regulation is hard to enforce and is often distorted by specific interests.

The previous discussion illustrates that the type of regulation has to be defined with respect to a specific field. This is what the workshop participants call sector-dependent regulation. In the case of e-commerce, it seems that the tools of general regulation (e.g. competition laws) are sufficient. On the other hand, there is a need for a specific regulation in the sectors that used to be organised around national monopolies and that are now open to competition; the critical mission for regulation in these sectors is to structure the market and create the conditions for competition. Furthermore, as already underlined by Jos Leyten, the choice of type of regulation has to be regularly reassessed: the general regulation of e-commerce is likely to persist, while the specific regulation in sectors in transition is of a more transient nature. There is thus a link between the sector-dependency and the time-dependency of regulation in the ICT sectors.

PART IV

Outlook

Pieter Ballon & Pascal Verhoest

The Belgian Information Economy: Contribution to a Federal Action Plan

Introduction

A fast-growing body of literature suggests the development of a knowledge economy. In such an economy, knowledge is an important, if not the most important factor of production. Opinions diverge as to the social consequences of such a change in the economic system and as to the meaning of the term information society which it is often associated with. However, most observers agree that the development and the implementation of information and communication technologies (ICTs) is to become a key element in determining the competitive strength of the economy. The development of the telecommunications sector is a crucial condition for this.

It is commonly assumed that in the development of the information economy, Belgium is lagging behind in comparison to most other western industrialised countries. This supposed shortcoming and some possible solutions are the subject of this article. Our analysis was originally devised as a critical and forward-looking supplement to the action plan for the information society of the Belgian Department of Economic Affairs and Telecommunications (BIPT, 1997)[1]. Whereas the official action plan was primarily oriented toward telecommunications policy, this article also takes into account issues related to industrial policy and innovation, in order to broaden its view to the whole information economy. However, it does not cover this subject exhaustively. The analysis focuses on the federal level of decision making. It is particularly inspired by an evident lack of co-ordination of research efforts subsidised by the countries' different governments (federal, Flemish, Walloon), which may be leading to wasteful duplication[2]. Our analysis of the Belgian information economy, based on earlier work done in this area (see e.g. Verhoest, 1996b), was placed in an international perspective, by drawing upon international literature, often published by supranational bodies such as the European Commission, the OECD and the UN.

The first part of this article presents a selection of the international statistical material available, and points to the strengths and weaknesses of the Belgian telecommunications sector.

The second part sets out some general guidelines that might inspire the government's telecommunications policy. One of the conclusions drawn is that the Belgian telecommunications sector is in need of a flexible and realistic institutional and regulatory framework. The final part of this article examines a number of other measures the federal government can take in addition to its telecommunications policy and that might stimulate the uptake of ICT in the Belgian economy and point to the institutional difficulties such a policy is likely to face.

1. Strengths and weaknesses of the Belgian telecommunications sector

In this part of the article we present an overview of recent developments in the Belgian telecommunications sector (1995-1998) by means of a comparative institutional and economic analysis. This overview takes four elements into consideration. Firstly, it presupposes that the development of the telecommunication sector correlates with the general state of the economy. The level of development of the Belgian economy, therefore, gives an indication of the *potential market* for telecommunication services. This can then be related to the actual level of demand for telecommunications services. This is the second element. In case of a divergence, additional factors need to be taken into consideration. Primarily, these are the quality and price levels of the supply in telecommunications services and infrastructure – the third element in the analysis. Finally, as the fourth determinant of market dynamics, we will consider the institutional and regulatory framework.

As mentioned in the introduction, we have brought together a set of international indicators to evaluate the position of the Belgian telecommunications market[3]. A basket of ten countries was created, containing countries with so-called proactive policies towards the telecommunications sector (the USA, the UK, Denmark, Sweden and the Netherlands) as well as countries with a more reactive attitude (Germany, France, Italy, Belgium). Portugal, a country characterised by major arrears, was included in the sample as well.

A warning is necessary because of the limited availability and reliability of international statistics. Moreover, the most recent comparable data are two to three years old, which is a long time considering the rapid developments in the market. For these reasons this analysis is limited to a number of basic indicators. Nevertheless, these supply a surprisingly clear image of the relatively international position of the Belgian telecommunications sector in recent years and of some possible causes of this situation.

1.1. Market potential

One indicator for the market potential of ICTs is the development of the technology-intensive industry and the services sector. The services sector is an intensive user of new information and communication technology, and is increasingly considered as the main driving force of ICT developments (Wyckoff, 1996). International statistics indicate that the share of highly technological industries in Belgium might be below that in other countries. The services sector, however, has a prominent place in the Belgian economy, accounting for 70,8 % of GDP in 1994. (OECD Observer: 200; OECD in Figures, 1996b).

Two other factors that could also influence the development and application of ICT are the size and the international scope of companies. It is assumed that small companies, with the exception of those in highly innovative sectors like the ICT sector itself, have much greater difficulties in implementing innovative technology than large ones. Companies with an international scope are more likely to be heavy users of new ICTs (OECD-TEP, 1992; SERV, 1996). In fact, Belgium has a relatively smaller number of large companies in comparison to countries like the Netherlands, Germany, the UK, Sweden and the USA (OECD, 1997a). As far as the international focus of the economy is concerned, some sources claim that no less than 70% of the Belgian GDP originates from exports (Van Bastelaer, 1997). Moreover, the strong presence of public, semi-public and private international organisations causes additional demand for cheap and innovative long-distance information and communication services.

All of this indicates that, along with some potentially unfavourable features, the Belgian economy possesses characteristics that are potentially favourable to the diffusion of technological (and organisational) innovations and applications of ICT. In particular, the highly developed services sector could exert a positive influence. In other words, there appear to be no significant reasons to assume that demand for telecommunication services in Belgium would seriously lag behind on that of the other countries in our sample. The next section will examine whether this supposition corresponds with reality.

1.2. Demand

In order to give an estimation of the demand for telecommunication services in Belgium, a number of indicators of the penetration of different telecommunications services are examined. It appears from table 1 that in most cases Belgium lags behind on the other countries in our sample. Sweden, Denmark, the United States, the UK and to a lesser extent, the Netherlands and Germany, are the leading countries. France, Italy and Portugal, together with Belgium, score badly on most indicators, except on ISDN penetration. The latter is explained, amongst other things, by technology-driven programs that date back to the period prior to liberalisation.

	Main lines per 100 inhabitants (1997)		Cellular mobile subscribers per 100 inhabitants (1997)		ISDN subscribers per 1000 inhabitants (1997)		Internethosts per 1000 inhabitants (incl. under .com, .net etc.) (july 1998)	
1	Sweden	68,0	Sweden	35,8	Germany	35,1	US	78,8
2	US	66,0	Denmark	27,5	Netherlands	17,8	Sweden	62,4
3	Denmark	63,6	Italy	20,5	Belgium	9,48	Denmark	51,3
4	France	57,6	US	20,4	Italy	5,85	Netherlands	42,1
5	Netherlands	56,6	Portugal	15,4	Portugal	4,88	UK	28,4
6	Germany	55,0	UK	14,3	US	3,68	Belgium	19,0
7	UK	54,0	Netherlands	10,8	Denmark	n.a.	Germany	18,3
8	Belgium	48,5	Germany	9,9	France	n.a.	France	11,4
9	Italy	44,9	France	9,8	Sweden	n.a.	Italy	8,0
10	Portugal	40,8	Belgium	9,6	UK	n.a.	Portugal	5,4

Source: OECD, 1999

Table 1. Demand for telecommunication and internet services

In general, however, table 2 shows that the use of telecommunications services in Belgium is below average compared to the other countries in our sample. This is also confirmed by other data, for example on the spending patterns of telecommunications users, and in which Belgium is placed at the bottom, or close to the bottom, of most categories considered (EITO, 1998). These observations contrast with our earlier estimate that there is sufficient potential in Belgium to support a flourishing telecommunications market. The question thus arises as to the determinants of these developments. As we have already mentioned, the data provided do not always reflect the latest developments in the telecommunications market. For instance, ever since a duopoly was introduced in mobile telephony, Belgium has been catching up in this area. This gives an indication of a positive influence of liberalisation measures, especially in profitable and fast growing markets such as mobile telephony. This issue is addressed in a following section. We first turn to some other possible element of explanation, i.e. the supply of services and the institutional framework.

1.3. Supply

The weakness of the Belgian telecommunication market might be explained if it appears that the supply of services and infrastructures is poor. Important aspects on the supply side of the telecommunications sector are infrastructure, the range of products and services, tariffs and the quality of service. As far as the telecommunication infrastructure and its (necessary) modernisation is concerned, the most widely used indicator is the rate of digitalisation of the voice telephony network.

With a percentage of 83 digital main lines in 1997, Belgium lags behind on all countries in our sample. The Netherlands, Germany, the UK and France all have a greater main line digitalisation. Moreover, the level of investments in Belgian public infrastructure in recent years is also below the average of the countries considered (OECD, 1999). Table 2 also shows that the tariffs of telecommunications and Internet services in Belgium are relatively high when compared to most other countries in our sample. Clearly, the Scandinavian countries have the lowest tariffs.

	Basket of business telephone charges (aug 1998)		Basket of residential telephone charges (aug 1998)		Basket of 1,5 / 2,0 Mbs leased line charges (aug 1998)		Peak rate Internet access basket (20 hours on-line per month) (1998)	
1	Denmark	468	Sweden	280	Denmark	408382	US	39,77
2	Sweden	473	Denmark	301	Sweden	416202	Italy	42,27
3	Netherlands	636	US	340	Netherlands	744101	Sweden	48,01
4	UK	826	UK	367	US	820204	Denmark	54,04
5	Belgium	908	Netherlands	376	Belgium	915473	Portugal	61,20
6	France	976	Germany	419	UK	919967	Netherlands	62,79
7	Germany	984	France	420	France	929630	Germany	68,44
8	US	1037	Italy	436	Portugal	1664608	UK	70,01
9	Italy	1067	Belgium	454	Italy	2111729	France	72,05
10	Portugal	1448	Portugal	660	Germany	n.a.	Belgium	97,70

Based on purchasing power parities, in USD. Source: OECD, 1999

Table 2. Charges for telecommunications services

With respect to quality of service, indicators include the waiting time for the installation of a new line, the waiting time for reparations, frequency of faults and itemised billing. The waiting time for installation by Belgacom, Belgium's incumbent operator, is four days. Ninety percent of all faults are repaired within 24 hours. Both scores are average, compared to our sample. Different criteria are used to record the frequency of faults, which renders a comparison practically impossible. Like most other OECD countries Belgacom now provides fully itemised bills to its customers (OECD, 1999).

It is nearly impossible to provide a statistical overview of the supply of new telecommunications services, and even more so to draw any international comparisons. In general, companies in Belgium seem to develop fewer new information and communication products and services, except for the financial sector. Reasons for this are amongst others the small scale of the market, high telecommunications tariffs and a lack of entrepreneurship (Lowette, 1995). The number of secure web servers gives an indication of the take-off of electronic commerce. In August 1998, there were 0,51 secure web servers for e-commerce in Belgium per 100,000 inhabitants. Again, this puts Belgium almost at the bottom of our sample. In contrast, the US had 6,13; Sweden 2,08; the UK 1,41; Denmark 1,01 and the Netherlands 0,95 (OECD, 1999).

In general we can conclude that, over the last few years, Belgium has held a relatively weak position as far as the supply side was concerned. Bottlenecks were amongst others relatively high tariffs, a lagging modernisation of the infrastructure and a limited supply of new information products and services. On the supply side, most indicators we compiled point to the Scandinavian countries, the US and the UK as having the best average supply of telecommunications services. Since many of these are countries that were at the forefront of the liberalisation of the telecommunications sector, this may point to the fact that other factors influence demand, i.e. institutional and regulatory factors.

1.4. Institutional factors

In order to render a comparison of institutional factors possible, we shall examine the efforts of various countries to install a favourable regulatory regime for the telecommunication sector. The European Commission, amongst others, has attempted to indicate what the objectives of such a regime are — the previously monopolistic telecommunications operators have to become independent players in a unified and liberalised telecommunications market. The 1987 European Green Paper made it clear that the ultimate goal of government regulation of the telecommunications sector had to be its eventual liberalisation. This happened in several stages. In 1988, competition was allowed in the area of terminating equipment. Two years later, directives for liberalising the so-called value-added telecommunication services were issued. The Open Network Provision put forward free access to telecommunication networks and the Bangemann Report of 1994 announced the liberalisation of infrastructures. Finally, on 1st January 1998, voice telephony was opened up to competition in most European countries.

In this process, a number of countries distinguished themselves by liberalising parts of the market faster and more efficiently. Most importantly, they prepared the former monopolist for a competitive environment in a proactive way. The whole process of preparing the sector for (full) competition usually meant starting to render the incumbent network operator more independent. More often than not this was achieved through privatisation. This ensured the operators a managerial structure enabling the company to operate like a commercial enterprise. This 'proto-liberalisation' generally went hand in hand with the gradual introduction of effective competition in order to allow the incumbent to become accustomed to a competitive environment. In contrast, other countries appeared to follow reactive policies and consequently found it hard to break with a number of options that belonged to the past.

During the years 1991-1994, Belgium clearly belonged to the latter category. The government opted for a minimalist interpretation of the European liberalisation guidelines. Also, the pace of liberalisation, set by the European Commission, was often deliberately slowed down. Officially, the Belgian government adopted this 'conservative' position out of concern for the weakness of the incumbent (known as Belgacom since 1991). A certain unwillingness to break with the system of politicisation undoubtedly reinforced this position. The system of politicisation (as understood in its broadest sense) also explains why the Belgian government was very reluctant to grant Belgacom more independence (Verhoest, 1996b).

Over the course of 1994 to 1998, the federal government gradually abandoned this position in favour of subordination to the European rules. Factors external as well as internal to Belgium explain this change. Amongst the external factors is the pressure from the European Commission as well as the growing threat of international competition. Internal factors are, for instance, increasing national competition among regional operators supported by the Flemish government such as Telenet, as well as the growing role of the regulatory authority BIPT (Belgian Institute for Posts and Telecommunications), founded in 1993. In turn, the independence of Belgacom was enlarged by the partial privatisation of the company in 1998. This privatisation didn't only serve the well-being of Belgacom, but certainly also the interests of the federal treasury. Along with this, privatisation was also accompanied by the imposition of significant social obligations, such as the foundation of its own pension fund and the maintenance of high employment levels.

It might have been more sensible to adopt a more offensive approach, like the one adopted in other small countries such as Denmark. These countries speeded up the independence of their incumbent operators as well as the introduction of regulatory pressures upon them. In this way, those countries that anticipated European liberalisation measures have acquired a strategic advantage over others. It is important to note that their operators are today not only well-positioned on the international market, but due, in part, to the stricter regulatory environment in which they had to operate, these countries have also managed to offer better and equitable prices and quality of service on the national market (see e.g. Verhoest, 1996b). This is the case, for instance, of Denmark a country with a population half as big as Belgium and whose national operator, TeleDenmark, is now shareholder of Belgacom).

The Belgian federal government has always legitimated its reactive attitude toward liberalisation by motives related to the 'national interest' or as measures serving social causes such as employment and universal service. It is, however, unclear what the costs of this operation will be in the long run and if they will result in a weakening of the company's competitive position. As argued, a guided liberalisation need not to be to the detriment of the national interest nor does need to be inequitable.

Table 3. Telecommunications regulatory regimes

	Voice Telephony		X.25 Data-Communication		Leased Lines		Mobile Telephony	
	1996	1994	1996	1994	1996	1994	1996	1994
Belgium	M 1998	M	C	M 1993	M 1997	M	D	M
Denmark	C juli 1996	M	C	M 1993	C	M	C	C
France	M 1998	M	C	M 1993	C	M	C	D
Germany	M 1998	M	C	C	C	M	C	D
Italy	M 1998	M	M 1998	M 1993	M 1998	M	D	D 1994
Netherlands	M 1997	M	C	M 1993	C	M	C	D 1994
Portugal	M 2000	M	C	C	M 2000	M	D	D
Zweden	C	C	C	C	C	C	C	C
UK	C	C	C	C	C	C	C	C
US aug 1996)	C (local partial C)	C (local C	C	C	C	C	C	

M=monopoly; D=duopoly; C=competition
Sources: OECD, 1995b; OECD, 1997b.

As the preparation of liberalising the market generally goes hand in hand with rendering the former monopolistic operator more efficient, the degree of liberalisation may serve as an indicator of proactive policy. Table 3 lists the different telecommunications regulatory regimes in the aforementioned countries for the years 1994 and 1996. The table indicates a direct relationship between institutional factors and the development of the telecommunications market. According to this table, liberalisation is highest in Sweden, the UK and the USA. Denmark is next. Germany and the Netherlands are in the middle of the pack, while France, Portugal, Belgium and Italy trail behind. This ranking closely corresponds with the development of the telecommunications market that was described earlier in this article.

It thus seems that there is a positive correlation between institutional factors, i.e. a proactive policy aimed at rendering the incumbent operator more independent and liberalising the telecommunication market, and the development of this market. Starting from this conclusion, the next main section turns to look at the origins of the problem and suggests some general conditions necessary for a successful federal telecommunications policy.

2. Telecommunications policy

An active and anticipatory policy aimed at the liberalisation of the telecommunication market appears to be an important condition for a well-developed telecommunication market. It can be expected that such a policy will remain of great importance in the future. The present opening up of the market to competitive forces does not mean that there will be no role for regulation. An 'idealist' scenario, in which a sufficiently large number of players have equal access to the market, is far off. A situation has arisen instead where the dominant party or parties are advantageously placed by having engaged in strong anti-competitive policies. If liberalisation is considered as a means to a higher end, i.e. the optimal development of the telecommunication market, then policy should counter anti-competitive strategies, even after the formal liberalisation of the market. In other words, liberalisation and competition are to be seen as dynamic processes, that make constant, 'proactive' monitoring and regulation necessary (Mansell, 1993; Melody, 1997c+d). Such policy will also have to perform the necessary social corrections if the consequences are deemed inequitable.

2.1. The importance of regulation

This can be illustrated on the level of the operational tasks of the regulatory authority, the Belgian Institute for Posts and Telecommunications (BIPT). This institute was set up as the regulatory body for the telecommunications sector in 1993, well after the establishment of similar institutes in many other countries. The operational tasks of the BIPT include crucial assignments such as handing out licenses, setting up and administrating both the numbering plan and the interconnection conditions. The case of interconnection illustrates the way in which the operational assignments of the BIPT influence the structure of the market. Due to the interdependence of telecommunication networks, efficient interconnection is a basic condition for competition. The setting of interconnection conditions and tariffs, therefore, has potentially a major impact on the development of the market. The lack of a tradition of detailed cost calculation and assignment, and the fact that interconnection requires co-operation and negotiations, presents opportunities for the dominant operator to abuse its market position (Melody, 1997c+d).

Of great importance to the level of the interconnection tariffs is the system of cost calculation. A great number of countries, including Belgium, are following the so-called 'historical model' of cost calculation. However, this model incorporates parts of the inefficiencies from the past into the present interconnection tariffs and indirectly into consumer prices. This can be avoided, for example, by switching to the so-called 'long-run average incremental cost' method, which only takes into account the 'real' present costs of additional output. This method is said to be forward-looking because it forces the operator to remain or to become efficient in the future, rather than to live off previous investments. Still, as the experience of the United Kingdom shows, the transfer to this method of cost calculation is a very gradual process, in which the regulator plays a crucial role in building up the necessary expertise in cost accounting and in assessing the impact of a new tariff structure upon the development of the market (CEC, 1997).

Proactive liberalisation is, therefore, not the same as blindly opening up the market. It not only creates new possibilities and perspectives, but uncertainties as well. Recent research confirms that a high level of uncertainty, especially in environments which require long-term investments, may have adverse effects (for non-competitive environments see: Henisz & Zelner, 1997). Consequently, potential investors need to have some faith in the political and institutional systems. They also need to be able to predict the behaviour of partners and competitors up to a certain extent. In other words, to create favourable conditions for the telecommunications market, it is necessary to find a balance between dynamics and trust.

2.2. Regionalism and internationalisation

In the context of the federalisation process in Belgium, the first area where uncertainty needs to be avoided is related to the 'regionalisation' of the telecommunications industry and policy, a trend which has steadily increased in recent years. Since a modern economy cannot function without telecommunications, and the regions are increasingly trying to get a grip on economic policy, this question has been of growing interest to them. The foundation of a 'proper' Flemish operator, Telenet, under the auspices of the Flemish government, is an expression of this trend. It is also expected that the regions will try to influence telecommunications regulation in some shape or form (Verhoest, 1994a).

Taking into account a further regionalisation of Belgium, without an actual falling apart of the country, it is advisable to use this development in a positive way. The obligation for Belgacom to provide uniform tariffs in both regions, for instance, might result in the short term in expanding the effect of competitive forces to both regions. Since the build-up of such regional operators generally requires international alliances (as was the case with Telenet), this may be a way to combine the paradoxical forces of internationalisation and regionalisation.

3.3. Institutional changes

A second matter requiring attention is the institutional framework. The impeding politicisation of the Belgian telecommunications sector has not been negated by the foundation of the BIPT. In addition to the aforementioned operational tasks, the BIPT also acts as an arbitration committee and prepares policy. This has been criticised in the past by, amongst others, the European Commission. The Belgian minister responsible for telecommunication has authority over both the regulator and the incumbent, of whom he also acts as the main shareholder. When the BIPT was given arbitration powers in 1995, the authority accumulated by the minister was threefold: the minister was now responsible for the incumbent, general telecommunication policy and arbitration. A change in the law at the end of 1997 partially deleted this arrangement. The new law gave the BIPT autonomy over a number of crucial arbitration functions, such as interconnection, equal access, and shared use of networks. Still, one can wonder if this separation of powers was rigorous or transparent enough (Verhoest, 1996a). In the present regulatory framework, the autonomy of the BIPT is still largely dependent upon the dynamics and the 'personality' of the minister and of the BIPT.

The phenomenon of convergence, i.e. the gradual converging of the telecommunications and the media sector into one big technological and economic system, adds an extra dimension to this problem, as it enlarges the need for an integrated telecommunication policy. Decisions concerning telecommunications (a federal matter) can influence the media sector (a competence of the Flemish and Walloon communities), and vice versa. Thus, the need for consultation between the federal level and the regional and community level will probably grow. This is another matter that needs to be taken into account by proactive policymakers. To summarise, a guided and equitable liberalisation of the market supported by a general institutional framework, aimed at bringing transparency and confidence, is the prime task of policy. Such policies should consider future developments on an institutional as well as on a technological and economic level.

3. Industrial and innovation policy

In the previous sections, we have described some general conditions necessary for a proactive policy in the telecommunication sector. However, policymakers can, and do take additional measures to obtain a well-developed information market. It is, however, necessary to ask whether this is desirable in a liberalised market that should be driven by demand. One can answer this question positively for the following reason.

The so-called information society is characterised by increasing economic internationalisation, which is in turn partly caused by developments in the information and communication sector. This leads to two observations. Firstly, the leading countries, which have mostly adopted a proactive liberalisation policy, are also taking additional measures to stimulate ICT development and uptake. Despite internationalisation, the support of national industries, to paraphrase Porter (1990), is still very much a part of the competitive advantage of a nation. Secondly, the changing 'technological and economic paradigm' which we are now experiencing, demands the 'fine-tuning' of a whole series of conditions and institutions. This seems to necessitate the active involvement of the government (for a critical review see: Ruigrok & Van Tulder, 1995). We will go through some of the major needs and suggest some possible solutions.

As noted in the introduction, we do not provide an extensive overview of all actions related to industrial and innovation policy that may be undertaken by the different governments and institutions in Belgium. Rather, we concentrate on the federal level of decision making and particularly on issues related to the co-ordination of (public) efforts in research and development.

Co-ordination becomes a key question in Belgium because of two developments: firstly, the regionalisation of Belgian policy, i.e. the division of competencies between different levels of government, and secondly, the increasingly networked character of the economy, which leads to an increase in uncertainty and interdependency of companies. Because knowledge-intensive activities tend to be characterised by increasing returns to scale the outcome of investment to ICT may depend on a critical mass of resources being devoted to it. The pooling of efforts has become a necessary condition for the development of the so-called information economy, but it is increasingly difficult to achieve

(Andersson, 1998: 11). The next sections deal with two co-ordination measures in particular: the development and use of foresight techniques and the stimulation of partnerships for innovation.

3.1. Foresight

Given the high level of uncertainty surrounding the development of a new economic paradigm it is important for governments to identify the key sectors in which investments are most needed. Major errors in identifying crucial developments in technology and the market can lead to a waste of resources (Freeman, 1987; Ferné, 1996: 95). This, along with other factors has brought foresight back as a tool for policy development[4]. There are some general indications already as to which key sectors may hasten the development and implementation of innovative ICT, such key areas being electronic commerce and government services (see e.g. Falch, 1997: 115). At the moment however, banking and insurance appear to be the only critical sector in which Belgium has a good innovation record. It must be added that the latter is the result of historical and geo-economic circumstances, not of a foreseeing and pro-active policy (Lowette, 1995).

Foresight can be defined here as a set of mapping and measurement tools and techniques that help organisations (companies, government, etc.) to define strategy and set priorities. At the same time, however, the methodology of foresight is in crisis as the traditional 'technology centrist' vision is increasingly replaced by an integrated approach that combines social and technological aspects (De Laat, 1997: 22-41). This means that, besides the existing methods of 'technology mapping', there is a need to develop foresight techniques that take into account institutional developments as well as user needs (Boden, 1992).

The usual methods to measure those needs are often inspired by marketing techniques and unsuitable to the tasks set. The usual research topic of these methods is the demand for a particular technology, thereby wrongfully presupposing that non-users have somehow acquired technical foreknowledge. Therefore, the real challenge of this kind of foresight analysis lies in the development of methods that chart the underlying information and communication needs of users. In addition, the process character of foresight techniques is becoming more important than the 'foretelling of the future' component (Metcalfe & Georghiou, 1998).

This can be seen in connection with the development of the networked economy. The network economy implies that innovations increasingly take place in interaction with a range of partners (customers, suppliers, regulators, knowledge providers etc.). Strategy formation therefore becomes partly a joint effort that is in the semi-public sphere. In this respect, foresight provides companies with information about their collaborators' and competitors' views on future developments. This reduces a firm's uncertainty about strategy and provides different firms with a common strategy for the future.

Foresight techniques that enlarge the policymakers' view of technological developments, R&D activities (both public and private) and market developments, may enable governments to create or strengthen the national 'knowledge infrastructure' (CENTRIM, 1998). A national 'knowledge infrastructure' points to a vision of national innovation organisations (both public and private) as a 'landscape of linkages' where governments must intervene with a systemic approach that is tailored to fit the characteristics of this infrastructure. Prioritisation of technologies and sectors consequently becomes possible. This then can steer the government's incentives policy.

3.2. Incentives

Once key sectors for economic and social development have been identified and focused upon, the question arises as to how the government can stimulate the development and implementation of ICT in an active way? Traditionally, public subsidies for the development of new technology would be given. According to this vision, market failure is most likely to happen in the primary, 'pre-competitive stage', and therefore, government intervention is only necessary at this stage. However, this kind of approach is increasingly being altered. The emphasis on stimulating the development of new technologies stems from a vision in which the development of a certain technology into a commercial product is linear. New insights that regard the development and the application of new technology as an interactive process, suggest a broader approach to stimulating innovation (OECD, 1995a: 91-97). Empirical findings that technology diffusion had a greater impact on productivity growth than direct R&D expenditures in the period 1970-1993 underlined this (OECD, 1997c: 17). Thus, the focus of government attention is gradually moving towards the process of diffusion of innovation (OECD, 1991: 24-25, 70-71). For instance, the focus of the European ESPRIT-programme (1994-1998) lies largely with the needs of users and the market. Emphasis is put on use and usability, meaning that stimulating easy access to information, services and technology is a key factor in this programme (CEC, 1997b).

There is also a shift in the policy concerning subsidies. The mere handing out of subsidies as a policy tool to stimulate technological development has increasingly been criticised. It appears that organisations receiving subsidies all too often use them for investments that had already been planned, and that in general, money from outside the organisation's own resources is spent with less accountability than would otherwise be the case. Considering the need to pay more attention to the diffusion of ICT and the relative inefficiency of simple subsidies, recent findings suggest that the traditional ways to stimulate technological development, such as handing out subsidies to private enterprise and the investment in public R&D-institutes and education, no longer suffice (Ministerie van Economische Zaken, 1995: 18). Instead, additional means of stimulation need to be taken into account.

Inefficiencies might be avoided by a clear identification of the needs of the private sector. For instance, it should be questioned whether there is a need to give large companies extra incentives for the implementation of ICT, since these have significant resources of their own. Thus, with respect to large companies there might not be a need for government to go beyond the general framework previously set out in this article.

However, many small and medium-sized enterprises (SMEs) do need extra incentives from the state. SMEs, which play an important role in the Belgian economy, often have a number of problems implementing ICT[5]. Prima facie evidence suggests that besides a lack of financial means to invest in innovation, the main problem facing SMEs is a lack of knowledge and information (of course, this argument does not apply to SMEs that are active in the ICT sector itself, those showing more often than not a bigger dynamic than large companies). Since the characteristic form of innovation in SMEs is the application of existing knowledge and techniques (Coehoorn, 1995: 2), this means a lack of external sources for information about technological and technical developments. These sources can be very diverse: universities and laboratories, research companies and consortiums, suppliers, customers etc.

The effectiveness of knowledge transfer depends upon the quality and quantity of the information that is offered, upon the quality and quantity of the ties between the SME and its sources, and upon the capacity of the SME to use the information in an effective way. Still, serious problems arise on all of these levels (OECD, 1993: 8; Coehoorn, 1995: 18-19). Recent research has suggested two alternatives to policymakers to assist them in tackling these problems: the creation of structures for the transfer of information, and the stimulation of networks and partnerships between SMEs, large companies, and research institutes.

The establishment of 'interfaces' or 'information brokers', to facilitate the transfer of relevant technological information between research institutions and SMEs, is motivated by the fact that direct links between such organisations often turn out to be ineffective. Reasons for this are on the one hand the cognitive and psychological barriers of the SMEs, and on the other hand a high abstraction level of knowledge and the specific role of research institutions, such as universities. Intermediate organisations such as call centres, technological centres or constancy agencies, that 'translate' the existing knowledge and know-how for SMEs, may help overcome the barriers to information transfer.

An important success factor seems to be that these information brokers respond to the actual needs and specificities of the SMEs, rather than merely packaging and distributing the information that is available (OECD, 1991; OECD, 1992; OECD, 1993; Coehoorn, 1995: 144-145; Charles & Howells, 1992: 92-93).

However, the first evaluations of international experiments with such interfaces show that they alone are insufficient to bridge the knowledge gap of SMEs. Serious problems can still arise when the state tries to transfer information to SMEs. For instance, experiences in France demonstrate that very few SMEs actually use the information provided by the state, even though they know of its availability. Among their reasons for preferring other sources of information was that the information offered by the state was not tailored to their demands, and more generally they were suspicious of state organisations (OECD, 1995: 78-84). An evaluation of Dutch innovation centres concluded that these 'information brokers' should play the role of connecting SMEs with the sources of information, rather than transferring the information themselves (Coehoorn, 1995: 147-148).

This leads us to the second way in which the state can foster the implementation of ICT in SMEs, that is, by stimulating links between SMEs, large enterprises and research institutions. The concept of networks and collaboration between economic agents is about 25 years old, but over the last few years, interest in the concept has once again been expressed. Networks originate as a means of protection against the insecurities of the market. They are drawn upon to perform tasks that are not routines conducted by the company. Networks can be defined as 'more or less complex systems for the exchange of information, horizontally or vertically'. They can be more or less formalised. Within a network, information is 'free' (OECD, 1993: 54-55). Experiences show that such networks can prove to be very effective. In practice, however, it appears that it is very difficult for small companies to enter a structured network and to obtain access to useful information. A lack of personal contacts seems to play a very important role in this, and one that is hard to influence (OECD, 1993: 64, 93-94).

In order to stimulate 'socio-technical' networks of SMEs, large enterprises and research bodies, government can encourage technological partnerships. Typical features of such partnerships are that[6]:

- they are mutually profitable and reflect the market interests of all involved partners. This excludes, in principle, those partnerships that are only motivated by subsidy programmes
- they include an explicit agreement to collaborate
- they can lead to lasting links between partners

- one of their central goals is to serve the learning process of all partners and to make a lasting contribution to the expansion of the technological capacities of every (but mainly of the weakest) partner
- they take place within a whole system of technological relationships and links.

Following the example of the ESPRIT programme, these partnerships can be broadened to include not only joint research projects, but also other kinds of activities (CEC, 1997b; see also Meyer-Krahmer, 1996: 225). In Flanders, the PLATO-project, which assigns managers from large companies as 'godfathers' to groups of SME managers, is an initiative that deserves to be expanded. In any case, the role of the state in such partnerships will have to focus on co-ordination and facilitation in addition to the provision of limited financial support[7].

The task of such a co-ordinator is primarily to make an inventory of existing initiatives and to indicate possible overlaps and complementarities. All too often there is a multitude of isolated and fragmented projects that do not generate the necessary critical mass. It is therefore necessary to co-ordinate efforts across ministerial departments and linguistic barriers (OECD, 1991: 24-25, 70-71). One reason for this is that networks and partnerships often have a cross-border dimension. In addition, economies of large scale have to be generated since there are often major investments involved with a high risk factor. In the Belgian context this means that linguistic boundaries are artificial in terms of companies and markets.

The intensive 'division of labour' in the federal Belgium renders overcoming these regional boundaries increasingly difficult for the authorities. In addition to that, on the level of the administration and organisation, sociological aspects come into play: the transfer of resources to other departments apparently decreases the legitimacy of the department. At a political level, an additional problem is that politicians generally desire concrete, short-term results, while ICT development programmes often yield results over the longer term. However, as economic competencies are transferred from the federal to the regional level, to provide the necessary co-ordination may become one of the prime vocations of the federal government.

Conclusion

There is a growing awareness that economic growth and innovation are processes in which many factors — technological, economic, social, institutional and cultural — affect one another. This article focused on economic and institutional aspects. The central question was how the federal government can influence the development of the national ICT market.

It was concluded that a pro-active policy in telecommunications, aimed at a thoughtful liberalisation of the market, seems to be a precondition. We emphasised the need for constant monitoring and regulation to obtain fair competition and a socially equitable market. It was also stated that policymakers should stimulate confidence by creating a transparent framework that takes into account future institutional and techno-economical developments.

A policy of liberalisation does not mean that discussion about which forms of competition are possible as well as desirable, becomes obsolete. The introduction of competition in a formerly protected market has a social price tag, and can lead (certainly in the short term) to a loss of employment. It appears, however, that the costs of prolonging the 'conservative' and 'protectionist' political mentalities from the past, outweigh the benefits. This is why policymakers search for ways in which the economic surplus created by competition can socially be capitalised upon. This means that policy should aim at turning the social costs of liberalisation into long-term positive effects.

Supporting the development of the market is one area where positive effects can be created. This is why policymakers need to take additional measures in the field of foresight and co-ordination. The introduction of competition in telecommunications goes hand in hand with new forms of international competition, but also with new forms of cross-border collaboration. Therefore, Belgian policymakers need to beware of economic localism. Boundaries that have a cultural and historical importance, do not necessarily make sense in economic terms. In this light, a federal Belgian policy for the information society that can stand the test of both regionalisation and internationalisation, calls once again for changes in political and economic mentality.

Notes

1 A summary of this report, which was commissioned by the Belgian Institute for Post and Telecommunications (BIPT) was published by Ballon & Verhoest (1998). The authors would like to thank Jason Whalley (TNO-STB for comments on an earlier draft of this article.

2 For example, one of the authors of this article who regularly acts as an independent assessor of research proposals submitted to the Flemish Government noticed that Dutch-speaking researchers are often not aware of relevant projects carried out by French-speaking researchers. It should be added that the author's 'cross-cultural' awareness was greatly enhanced by participating in the research of the Interuniversitary project 'Research & interdisciplinary evaluation of the information society networks, usage and role of the state', funded by the federal institute for science policy (DWTC). Partners are CRID-CITA (Fundp), LENTIC (Ulg) and SMIT (VUB)

3 Not all of the indicators used in the original study are included in this article. We limit ourselves here to the most common ones.

4 Other factors in the growing popularity of foresight are the shrinking resources of the state, which has favoured the systematisation and prioritisation of science and technology budgets, and the growing influence of business on those policies (Grupp & Linstone, 1999).

[5] For an account of the concrete implementation of ICT by Belgian SMEs, see Henrotte a.o., 1996; Dolfeyn & Meert, 1996.

[6] This definition is derived from literature about partnerships between companies in the so-called developed and developing countries but can also be applied, mutatis mutandis, to partnerships between companies in the western world (UNCTAD, 1996c: 5).

[7] (CEC, 1991a; Alabau, 1997). A study into the economical effects of strategic technological partnerships found no direct link with economical profitability, but it did find an indirect one. Companies which are R&D and innovation-intensive have in general a high degree of profitability, and are also among those that engage most in partnerships (CEC, 1991b: 5). Another study confirmed that this last conclusion is even more true in the services sector (Poot a.o., 1997: 14-16).

Jean-Claude Burgelman

Innovation of Communication Technologies: some general lessons for the future from the past

Changes in technology not only open up the debate, as discussed in the previous pages, on what telecommunications policy to follow in the future, but also raises questions with regard to what innovation policy is most adapted to the challenges outlined. In particular because of the nature of these technologies that after all dealing with human processes, traditional push strategies with regard to innovation in this area cannot be copied.

In line with the work of the IUAP and in view of future work, SMIT wants to conduct as to public policy for innovation can be outlined in this area, some reflections on trajectories of innovation in information and communication technologies are presented here. Past experiences in innovation, implementation and market penetration of communication technologies suggest four kinds of hypothesis (Flichy, 1991; Burgelman, 1995c):

- firstly, technological and social development are often at a tangent with each other. It is not because the technological capacity is readily available that it will or shall be applied
- secondly, what is technologically feasible does not always take place, even though it is often represented as such
- thirdly, a great number of non-technological aspects determine the success — i.e. wide distribution and/or market penetration — of communication technologies
- and fourthly, communication technology trends can only be forecast to a certain extent. Valuable forecasts are forecasts put in a negative wording: i.e. that the extent to which a certain innovation might fail.

1. Innovation of communication technologies: more than just technical know-how

Thinking about a technology policy in general and a communication-technology-related innovation in particular, is to a great extent characterised by extreme technological determinism or technological 'bluff' (Ellul, 1988). Hereby it is taken for granted that what is technologically feasible shall occur socially as well. It implies that the social application or impact is induced from the technical make-up of the innovation. This deterministic reasoning isolates one societal phenomenon, which in its turn is readily accepted as an explanation for other developments.

The technological impetus to the development of communication technologies is as such not related to the social forces actually moulding it — albeit to a certain extent. This generates a tendency towards accepting the technical innovative capacities as sufficient and/or necessary conditions for social change. Using a witticism, a typical technologically deterministic reasoning would read as follows: 'more bandwidth inevitably leads to more communication and hence a better life'. Put differently, if one wants to realise an information society — of which it is assumed that it will have a higher level of civilisation than the industrial era — one should provide bandwidth, i.e. information highways. This type of technological determinism triggers off information technology 'hypes': i.e. each kind of communication technology which transmits certain sorts of communication (e.g. graphics, sound, text, data, etc) in far more better ways than the current technologies is in principle regarded as highly successful.

The CD-I was once assumed to be in great demand. But at the end of 1997, after years of fierce marketing, Philips decided to terminate the project, and even took the opportunity to shut down the whole multimedia division.

In the mid-eighties Videotext was, by the same token, expected to become a new mass medium. Almost all European countries invested massively in Videotext. At present Videotext has — at the cost of several ECU billions — reached a dead end all throughout Europe. Most forecasts were actually very wrong (more than 100%, see Burgelman, 1989).

One exception to the rule, though is France where France Telecom — the initiator of Minitel — allows their technological engineering to go hand in hand with a lot of social engineering. Indeed, Minitel was freely distributed, the users could easily and cheaply check the possible benefits, etc.

The videodisk was, in the same way, forecast as a medium that would push its competitor's VCR (Sony) out of the market. But it turned out to be otherwise: the videodisk disappeared together with the RCA parent company, which had it developed at an extremely high cost.

Similarly, launched in the fifties and sixties, cable was regarded in the same way as Internet is seen nowadays: a technical channel, which leads to some social (r)evolution. The social revolution didn't come, but something did change, though; the number of TV channels was multiplied by a very significant factor.

In short, technological know-how or technological superiority is not the sole explanation, but only one of the factors that determine whether a certain innovation breaks through or not. The last example is as such a real case in point:

at the end of the eighties, one book was often discussed within policy circles in general and EU seminars in particular, i.e. H. Ungerer's Télécommunications en Europe. Ungerer was then seen as one of the most important 'brains' in matters of telecommunication and DGXIII (responsible for telecommunications and information technologies) liberalisation. The author predicted, on the basis of the ultrafast development of microelectronics, that the number of available communication technologies would have multiplied exponentially by the beginning of the next millennium. A neat comparison with the development of car technology was used to illustrate the breathtaking technological advances in the PC world. If the car would have developed in the same way as the PC, an average car would now only weigh some good 5 kilos, have a top speed of 5000 km/h, consume one thousandth of one millilitre per kilometre of fuel and sell at a ratio of 10 billion cars a year. The book — prefaced by Delors and the then General Director, M. Carpentier — as such predicted then numerous intelligent applications of telephone, cable, etc. Outline (Ungerer, 1988; 22).

Telecommunications: Prospects for 2000

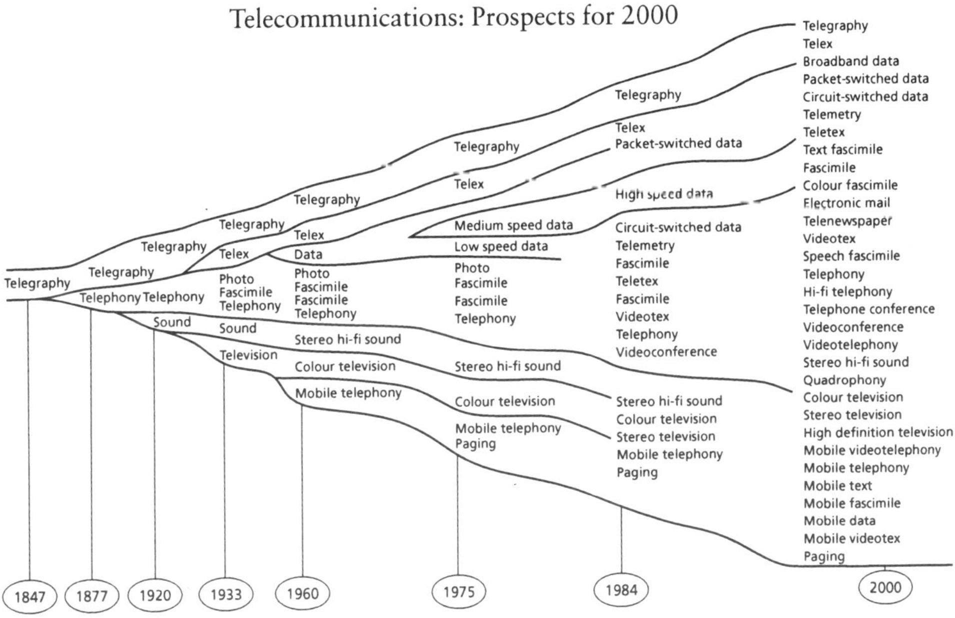

Source: CEC study

Analysing this forecast teaches us that this well-trained and influential top-ranking DG XIII civil servant — daily dealing with innovation in communication technologies — assessed almost everything wrongly. The striking thing is of course that one technology, currently hyped and very much a success as well, was not foreseen at all, i.e. Internet and its possibilities for convergence. This omission of one of the most important communication technology-related innovations shows that a rational extrapolation on the basis of some technological paradigms only, simplifies the whole business of innovation. Indeed, Internet is a type of technology that became an overnight success by sheer accident. But again not merely accidental, if one accepts that only with the introduction of WWW (which triggered off the social factor of user-friendliness and due to the lower telecommunication costs, the economic factor of accessibility) Internet was allowed to take full shape. It should also be added that the current Internet success preceded by 20 years of (military) governmental subsidies and academic R&D. This only goes to show that — as opposed to the presently rather dominant views stating that the government does not play a vital role in the field of innovation — a governmental policy is essential. Could the same time and energy indeed be invested in the 'inventing' of the Web within a fully liberalised market as we know today? An overall competitive environment without governmental support, logically tends to lead to a kind of innovation culture focused on short-term return-on-investments. And all major breakthroughs within the area of communication technology innovations are characterised by precisely the opposite, i.e. a long-term view or vision.

In a word, if one wants to assess future developments as correctly as possible then, — as shown by this brief introduction — it does not suffice to have only an eye for the technological capacities of certain innovations. After all, communication technologies are just the carriers of communication processes. The latter are exclusively human affairs and therefore determined by social, economic(al) and political incentives for people. Taking all this into consideration and within the limited scope of this article, I want to answer the following question: which lessons can be learned from the non-technical study of previously failed or successful innovations within the communication sector.

2. A full substitution of communication technologies and/or a sudden boost in communication services are rare

What is meant is that — as opposed to the current views — certain communication technologies are rarely fully substituted by other ones. By the same token, the TV industry has not ousted the film industry in contrast to what had been endlessly predicted. On the contrary, the film industry is currently one of the most important agents in the multimedia business. Likewise, e-mail or electronic communication in general shall not supplant the post. On the contrary, a kind of complementary relationship between the two channels. is taking shape. In the same way, the massive introduction of the PC has not led to the disappearance of printed paper or the breakthrough of the paperless office.

This general rule of complementary relationships instead of substitutions is an exception as far as innovation in the nature of process or procedure innovation is concerned. It is important to note that with this type of process or procedure innovation, the innovation cycles tend to be rather short term — 10 to 15 years before they reach full substitution — and that the market suffices as an ordening principle to support the innovation.

On the other hand, a radical innovation — in the neo-Schumpeterian sense of the word — is an innovation introducing new modes of operation, generating new social relationships and taking place in some technological cluster. Some examples: the engine or Internet. Such an innovation is often much more laborious; it takes longer (20 to 30 years) and as such necessitates a framework policy.

The striking thing is that the policy discussions on communication technology-related innovation often mix both types of innovation. It was initially expected that Telenet would introduce a neo-Schumpeterian innovation paving the way for Flanders to IS, boosting employment etc. But the policy was typified by its absence in favour of the market. Which in its turn explains why eventually Telenet became nothing more but an alternative operator and Internet Service Provider.

3. Emergence within the development of communication technologies

If one studies the greater breakthroughs in the field of communication technologies (such as telephone, radio, Internet, etc.) it is noticeable that these radical innovations are marked by a high degree of emergence.

I have borrowed this concept from evolutionary biology. It refers to the fact that the total is more than just the sum of its parts.

'Things can have certain qualities in combination, qualities which they do not have taken separately. The additional features are derived from the osmosis of the separate parts. We call them 'emergent qualities'. (...) The flying of birds was probably the result of the simple invention of applying feathers for insulation. We cannot always predict what will happen. Most new acquisitions in evolution are probably the result of these emergent qualities.' (Calvin, 1992: 109)

Revolutions within the communication technology sector are indeed characterised by a very high degree of emergence. This implies, that depending on the context — which is socio-economically determined and varies through time and space — some unexpected applications of communication technologies will crop up which of course do have to be potentially and technologically inherently possible. That is why different and unexpected applications of certain technologies break through, rather than the ones which had been hoped for within a purely technical ideology.

The invention of the telegraph was actually an accidental spin-off during the search for more efficient railroad management. Radio, for instance, was discovered when researchers got side-tracked in their search for a better telegraph system. Internet is likewise the spin-off of the search for efficient and the least vulnerable military communication. These 'spin-offs' are non-intentional and consequently not 'plannable' either, but each time, they can be explained in terms of a technical emergence within the current socio-economic dynamics.

Indeed, the technical know-how of telegraphing became widely available as soon as the stock exchange (as a new economic tool for ever-developing 19th century capitalism) became fully operational. This created a need for faster and more efficient information exchange in order to reach speculative objectives. The development of the radio, for example, took place when the new industrial bourgeoisie established themselves during the interbellum period and the mass-marketing of cultural consumerism. Theatre seats were limited — since they had been developed for the upper classes in the previous century — and radio was a means to eavesdrop from a distance.

Finally, Internet met the demand of the academic world for cheap communication for research purposes. But it also complied with the need for individualism with youngsters during the eighties and nineties and related to a new youth culture and general feeling of solidarity. Virtual Communities are after all nothing more than just that: being on your own in a group. It is therefore quite striking, but not strange, to notice that many of the current top-ranking industrial people within the internet area, used to belong to the 'alternative' scene at the end of the seventies.

This emergence principle is very important since it implies that anyhow radical changes can hardly be predicted at all.

4. Hype is a bad advisor

Another lesson to be learned from the analysis of past successes and failures is that the technological hype pervading the field of new communication technologies and cropping up every time certain technologies provide the means to bridge time and space, is often far more removed from actual applications.

Indeed, the discourse in the past upon the introduction of telegraph, telephone, radio, television and Minitel has to a great extent been similar to the discourse on the information highway: i.e. revolutionary changes are expected, drastic social changes are foreseen and spectacular breakthroughs in solving social problems are hoped for.

A similar 'technology push' vision is not only slightly simplistic — social changes are never due to one single reason — but also shows a fundamental lack of insight into user dynamics: i.e. what is the ultimate motivation for the user to apply media or communication technologies and how does the user operate.

5. The user as the great unknown

New developments are rarely seen or thought about from the point of view of the user: how and within which limitations does the user acquire some benefit from what happens (Burgelman, Punie & Verhoest, 1995b, chapter 3). If there is none, what should be done about it to make it happen? Using television as an example again: whatever new technology, when the user switches on the television set he/she shall want the programmes to satisfy his/her educational, informative or escapist needs the way he/she is used to. Digital TV or not, to the consumer it remains television and that is the way the user discovers the new media.

And if the user's point of view is taken into account, comparative data are not often simultaneously processed. Which means that, when one enquires after the success, acceptance and use of information technology within the professional field, it cannot be easily transposed to the possible success, acceptance and use within the domestic or private areas.

The social and/or economic added value and/or relevance of new communication technologies for/to the user differs fundamentally depending on whether they are applied within a professional or domestic area. This relationship added value/relevance is called for by analogy with a concept typical of the world of publishing, 'acquisition substitution' (Van Den Brink, 1987). It implies that the user will accept a new product more easily when it replaces an old one, with as a result a higher quality and/or lower price.

Available research allows one to conclude that the acceptance of new communication technologies within a professional area is highly dependent on the fact whether the technology will decrease the cost of communication (e.g. mailboxes) and/or will increase the quality of communication (e.g. intranet). In general, one rules applies: i.e. each piece of new communication technology that boosts the return-on-investment at the lowest possible cost, has the highest chances of success. This is, in its turn, dependent on the number of variables inherent to the implementation process of new technologies within the labour process in general.

Plugging into a financial database is — despite its cost — much more efficient than following the same transactions in the newspapers. Non-rational elements or considerations of status can play an important role though (e.g. the success of the car telephone) or just the need for escapism or social communication. The fact that in spite of the higher cost (no air fare) and more human effort (transport, away from home, etc.), personal contact is seen as much more worthwhile and pleasurable can explain the failure of videoconferencing.

One should also take into account that new communication technologies will only be accepted within the domestic area, when they provide existing communication processes at a lower price (e.g. faxes vs post) and/or of better quality (e.g. CD vs vinyl). Cable television is a good case in point. It complies with both: the quality of the television image is much higher than through hertz distribution, while the user cost is not significantly higher (i.e. divided over the years).

Acquisition substitution also plays a role in the domestic area but only in relation to recreational, social, familial and communicative applications. In other words, only applications that add to the level in which communication complies with the 'sociability' as a citizen or family member. So, one should not advertise ISDN to the everyday user arguing that it has so far unknown possibilities. Because, on the one hand the everyday user is just not interested and on the other hand nobody needs these possibilities to enhance social communication. As far as domestic use is concerned, it is illustrative to remind one of the fact that the most successful innovation in the field of communication technology ever since World War II happens to be video. This was actually quite accidental. In the sixties everybody was convinced that the Video Disc Player (i.e. the audio-visual compact disc) would beat its competitor, the video recorder, because it was technologically speaking more innovative and performant. Which was correct in its own right: VDP was more advanced. But one tended to neglect the fact that the use of video enabled a new kind of social praxis within the user's communication patterns and the user tended to see it that way as well. Video allowed the general public to have the so far impossible 'viewer's autonomy'. More in general, video provided the consumer with unlimited possibilities as far as time management was concerned. Or in a more abstract sense: video provided consultancy facilities to a previously purely allocationary medium. This was the video recorder's major trump which nobody had predicted, and against which VDP did not stand a chance.

Because of the increasing importance of innumerable communication and information processes on the one hand and the high expenditure on research and development within the field of new communication technologies on the other hand, it is quite self-evident that both the social and commercial policies will benefit from accurate insight into the different parameters determining the success (acceptance and use) of these technologies according to a user's point of view.

In short, consumer research has shown us that knowledge about the user tends to be limited and that some innovations are often designed to meet the needs and logic of a few people whereas most products are consumed by people whose logic is determined by a socio-economic reality which is at a tangent with the designer's. The fact that the greatest part of the consumers' purchasing power is situated within the greying and ageing part of the population and that exactly this part of the population tends to be the most technophobic, supersedes over the scenarios of the information society.

6. Conclusion

From the foregoing, one can logically infer that any innovation policy in relation to communication technologies and hence information highways, should have a keen eye to a non-technical approach. This implies controlling social practices and economic realities supporting the communication process.

However, as indicated in the paragraph on emergence, such an approach should not lead to some sort of user dictatorship or some user determinism. Because of the simple reason that it cannot be assessed either whether something succeeds or not, and also because the user is just not a structural actor within the innovation process. Still, by taking the user into account, failures can be avoided to the greatest possible extent.

Especially in order to avoid user determinism or immobilism, any innovation and related research into communication technologies should be focused on enabling conditions and platforms (Burgelman & Van de Perre, 1995a). But also in that particular case one should point out that platforms and conditions can only be set up and be successful when the foundations are present. As long as the number of modems for domestic use remains relatively low in Belgium, the relevant critical mass will never be reached. The same goes for the prices of Internet connections: as long as they are relatively high, critical mass won't be reached either. And that's where innovation policy and telecommunications policy meet. Large-scale innovation will indeed also require a large-scale infrastructural approach towards the end-user.

Paving the way for a new telecommunications policy implies laying the foundations for an integrated innovation policy.

References:

Alabau, A. (1997), Telecommunications and the Information Society in European Regions, *Telecommunications Policy*, Vol. 21, n° 8, pp.761-771.

Analysys Ltd. (1995), The costs, benefits and funding of universal service in the UK, *Final Report (95200) prepared for OFTEL*, Cambridge and London.

Analysys Ltd. (1997), The Future of Universal Service in Telecommunications in Europe, *Final Report for European Commission DGCIII/A1*, Cambridge.

Andersson, T. (1998), Managing a systems approach to technology and innovation policy, *OECD STI Review 22*, pp. 9-29.

Ariès, G. & Duby, G. (red.) (1990), *Histoire de la vie privée*. Tome 5. Seuil, Paris.

Ballon, P. & Verhoest, P. (1997), België en de informatiemaatschappij: Bijdrage aan een economisch en institutioneel actieplan'. *Study commissioned by the Belgisch Instituut voor Postdiensten en Telecommunicatie*. (Belgium and the information society)

Ballon, P. & Verhoest, P. (1998), Bijdrage aan een Belgisch actieplan voor de informatiemaatschappij, *Economisch en Sociaal Tijdschrift*, LII, 4, pp. 509-534. (Contribution to the Belgian action plan for the information society)

Bangemann, M. et alii (1994), *Europe and the Global Information Society. Recommendations to the European Council*, CD-84-94-290-EN-C, Brussels, May 26th. Available at http://www2.echo.lu/eudocs/en/report.html.

Bangemann, Martin et alii (1994), *Europe and the Global Information Society. Recommendations to the European Council*, CD-84-94-290-EN-C, Brussels, May 26 1994. Available at http://www2.echo.lu/eudocs/en/report.html.

Bannock, G., Baxter, R.E. and Davis, E. 1992 (1972), *The Penguin Dictionary of Economics*, fifth edition, Penguin Books, London.

Barry, N. (1997), Conservative Thought and the Welfare State. *Political Studies*. Vol. 45 n°2, pp. 331-345.

Bastelaer, B. van & Lobet-Maris, C. (1999) 'Multimedia in France', SLIM National Report, in Williams, R. and Slack, R. (eds) 'Europe Appropriates Multimedia. A study of the national uptake of multimedia in 8 European Countries and Japan', University of Trondheim, Senter for Teknologi of Samfunn (STS), report n° 42. pp. 123-156
http://www.info.fundp.ac.be/~cita/slim/France.html.

Bastelaer, B. van et al. (1997), 'The development of Multimedia in Belgium'. *SLIM National Report.*

Baumol, W. J., & Sydak, G. J. (1994), *Toward Competition in Local Telephony*. The MIT Press, Cambridge.

Becker, J. (1995), *Information for all or knowledge for the elite? The contours of a dissimilar European information policy*', Prometheus, Vol. 13, n° 1, June.

Belgacom (1997), *Belgacom's Reference Interconnect Offer for Public Network Operators*. Brussels.

Bianchi, A.F. & Richeri, G. (eds) (1996), *Telecommunications; new dynamics & driving forces*. IOS press, Amsterdam.

BIPT (1997), *La société de l'information: Note d' orientation et d' action*. Brussels.

Blackman, C. R. (1995), 'Universal service: obligation or opportunity?', *Telecommunications Policy*, Vol. 19 n° 3, pp. 171-6.

Blackman, C.R. (1998), Convergence between telecommunications and other media. How should regulation adapt?, *Telecommunications Policy*, Vol. 22 n° 3, pp.163-170.

Board of Trade (1994), Creating the Superhighways of the Future: Developing Broadband Communications in the UK, presented to Parliament by the President of Board of Trade by Command of her Majesty, November.

Boden, M. (1992) *The Identification of Technology Priorities for European Research and Technology Development: A Review of Technology Mapping and Related Techniques*. Monitor-Sast Activity.: CEC, Brussels & Luxembourg.

Bouwman, H. & Janowski, N. (eds), *Interactieve media op komst*. Otto Cramwinckel, Amsterdam. (Upcoming interactive media).

Breton, Th. (1994) *Les téléservices en France. Quels marchés pour les autoroutes de l'information?*, La Documentation Française, Paris.

Brosveet, J. & Sørensen, K.: (1999), *Multimedia in Norway: An Overview*, SLIM National Report, in Williams, R. & Slack, R. (eds): *Europe Appropriates Multimedia. A study of the national uptake of multimedia in 8 European Countries and Japan*, University of Trondheim, Senter for Teknologi of Samfunn (STS), rapport n° 42, pp. 333-366. Also available at: http://www.ntnu.no/sts/SLIM/slimpaper.html.

Browning, J. (1994), Universal Service: An Idea Whose Time Is Past'. *Wired*, n° 11, pp. 102-154.

Bruno, M. and Pleskovic, B. (eds) (1996), *Annual World Bank Conference on Development Economics*, Washington DC.

Burgelman, J.C. & Pauwels, C. (1992a), *Audio-visual and cultural policies in the small European countries: the challenge of an unified European television market*, Media, Culture and Society, Vol.14, pp. 169-183.

Burgelman, J.C. & Van de Perre, G. (1995a), De Vlaamse Informatie Samenleving (VIS) en de gebruikers, draft text for the subcommittee 'Information society & Users' of the Flemish Council for Science Policy. Brussels. (The Flemish Information Society and the Users).

Burgelman, J.C. & Verhoest, P. (1996b), Convergence and the Trans-European Information networks: a problematic EC policy. pp. 7-18 in Bianchi, A.F. & Richeri, G. (eds): *Telecommunications; new dynamics & driving forces*. IOS press, Amsterdam.

Burgelman, J.C. (1989), Voorspelde en te verwachten ontwikkelingen van interactieve media (Forecasted and expected developments of interactive media), p. 145-154 in Bouwman, H. and Janowski, N. (eds): *Interactieve media op komst*. Otto Cramwinckel, Amsterdam. (Upcoming interactive media).

Burgelman, J.C., Punie, Y. & Verhoest, P.: (1995b), *Van Telegraaf naar Telenet. Naar een nieuw communicatiebeleid in België?*, VUBPRESS, Brussels. (From telegraph to telenet. Towards a new communication policy in Belgium?).

Burgelman, J.C. (1992b), Van massacommunicatie naar communicatiepatronen (From mass communication to communication patterns). pp. 63-82 in Bardoel, J. and Bierhoff (ed.): *Media theorie*. Nijhoff, Amsterdam.

Burgelman, J.C. (1995c), Assessing information technologies in the information society: the possible relevance of communication science and research in A. Calabrese, S. Splichal & C. Sparks (eds) *Information Society and Civil Society. Contemporary perspectives on the changing World Order*, Purdue University Press, West Lafayette, pp. 185-207.

Burgelman, J.C. (1996a), *Uitdagingen voor het Communicatiebeleid, Informatie en Informatiebeleid*, Vol. 14 n° 3, fall, pp. 51-61 (challenges for communication policy)

Burgelman, J.C. (1997a), Issues and assumptions in communications policy and research in Western Europe: a critical analysis in P. Schlesinger, R. Silverstone & J. Corner (eds.): *International Media Research. A critical Survey*. Routledge, London & New York, pp. 123-153.

Burgelman, J.C.: (1997b), Service universel, service public et souci de diversité. Le débat sur les autoroutes de l'information', in *Réseaux*, n° 78: pp. 41-52.

Burgelman, J.C. (1997c), Communication Policies for Access. ASCOR launching conference Communication for Open Societies. Amsterdam, 18-19 September. Also presented at the CTP section "Access, Policies and exclusion", of the IAMCR Conference in Glasgow - 30 July 1998.

Burgelman, J.C. (1998), Information Society: changing rules of the game, impacts and new policies was held at the Institute for Prospective Technology Studies (DGXIII), Seville, April 98. Also presented at the EURO CPR conference in Venice, 21 March 1998, and published in the conference proceedings as *Communication policy and its challenges for the welfare state*, Communication & Strategies, n° 30, pp. 289-297.

Burgelman, J.C. (1999), Communication Policy and its Challenges for the Welfare State. In Calabrese, A. & Burgelman, J.C. (eds): *Communication, Citizenship and social Policy: rethinking the limits of the welfare state*, Rowman & Littlefield, Boulder.

Calabrese A., Splichal S. & Sparks C. (eds) (1995), *Information Society and Civil Society; Contemporary Perspectives on the Changing World Order*. Purdue UP, West Lafayette.

Calabrese, A. (1997), *Creative Destruction? From Welfare State to the Global Information Society*, Paper presented at the EURICOMM/SMIT Conference; 'Communication, Citizenship & Social Policy: Rethinking the limits of the Welfare state'. Boulder (Colorado), 2-4/10.

Calabrese, A. & Burgelman, J-C. (eds) (1999), *Communication, Citizenship, and Social Policy, Rethinking the Limits of the Welfare State*, Rowman & Littlefield, Boulder.

Calvin, W.H. (1992), *De rivier die tegen de berg opstroomt*, Bert Bakker, Amsterdam. (The river that flows up the mountain).

Castells, M. (1989), *The Informational City: Information technology, economic restructuring, and the urban-regional process*, Basil Blackwell, Cambridge.

Castells, M. (1996), *The Rise of the Network Society - The Information Age: Economy, Society and Culture*, Vol. 1., Blackwell, Oxford.

Castells, M. (1997), *The Power of Identity - The Information Age: Economy, Society and Culture*, Vol. 2, Blackwell, Oxford.

Castells, M. (1999), 'The Informational City is a Dual City', in Schön D., Sanyal B. & Mitchell W: *High Technology and Low-Income Communities*, MIT Press, Cambridge MA.

Castles F. & Mitchell D. (1992), *Identifying Welfare State Regimes*, Governance, Vol. 5 n° 1, pp.1-26.

CEC (1991a), *Key Factors for Industrial Partnership in EC Programmes*. Monitor/Spear. Office for Official Publications of the European Communities, Luxembourg.

CEC (1991b), *The Economic Effects of Strategic Partnerships and Technology Cooperation*'. Monitor/Spear. Office for Official Publications of the European Communities, Luxembourg.

CEC (1996), *Information Technology: Solutions for Business*. Office for Official Publications of the European Communities, Luxembourg.

CEC (1997a), *Commission Recommendation on Interconnection in a liberalised telecommunications market*. Part 1- Interconnection Pricing. C(97) 3148.

CEC (1997b), *The deployment of the information society for regional development: Examples of applications from European Commission actions*. Office for Official Publications of the European Communities, Luxembourg.

CENTRIM (1998), 'RTOs in the service economy: knowledge infrastructures, innovation intermediaires and institutional change'. Proposal to the Commission of the European Communities, DG-XII.

Cerny, Ph. (1990), *The Changing Architecture of Politics*, Sage, London.

Cerny, Ph. (1991), *The limits of Deregulation: transnational interpenetrations and policy change*, European Journal of Political Research, Vol. 19, pp. 173-196.

Charles, D. & Howells, J. (1992), *Technology Transfer in Europe: Public and Private Networks*. Belhaven Press, London & New York.

Coehoorn, C. (1995), *The Dutch Innovation Centres: Implementation of Technology or Facilitation of Small Enterprises?*. Labyrint Publication. Capelle a/d Ijssel.

Collins, R. & Murroni, C. (1996), *New Media Policies*. Polity, London.

Cronin, F.J. e.a. (1993), *Relative demand for telecommunications", Information Economics and Policy*, V, 1.

Cuilenburg, J. van (1997), 'Access and diversity in communications and information', ASCOR launching conference 'Communication for Open Societies. Amsterdam. 18-19 September.

Cuilenburg, J. van and P. Slaa (1994), 'Competition in the Local Loop: Towards an Anticyclical Competition Policy in Telecommunications', in: *Telecommunications Policy*, Vol. 18 n° 1, pp. 51-65.

Cuilenburg, J. van and P. Verhoest (1998c), 'Free and Equal Access to Communications: In Search of Policy Models for Converging Communications Systems', in: *Telecommunications Policy*, Vol. 22, n° 3, pp. 171-181.

Cuilenburg, J. van and Slaa, P. (1993), 'From Media Policy towards a National Communications Policy', *European Journal of Communication*, Vol. 8, n° 2, pp. 149-176.

Cuilenburg, J. van (1998a), 'Diversity Revisited: Towards a Critical-Rational Approach to Media Performance, in: K. Brants, J. Hermes and L. van Zoonen, (eds.): *The Media in Question*, Sage Publications, London, pp. 38-49.

Cuilenburg, J. van (1998b), 'New Perspectives on Media Diversity: Toward a Critical-Rational Approach to Media Performance', in: Y.N. Zassoursky and E. Vartanova (eds.): *Changing Media and Communications*, Faculty of Journalism / Publisher ICAR, Moscow, pp.71-85.

Culpit, I. (1992), *Welfare and Citizenship, Beyond the Crisis of the Welfare State*, Sage, London.

Curtil, C. (1996), *La carte française des inforoutes*, Hermès, Paris.

Dahlgren, P. & Sparks, C. (eds.) (1993), *Communication and Citizenship. Journalism and the Public Sphere*. Routledge, London.

Davies, A. (1994), *Telecommunications and Politics: The Decentralised Alternative*. Pinter Publishers, London.

De Laat, B. (1997), *Scripts for the future: Technology foresight, strategic evaluation and socio-technical networks: the confrontation of script-based scenarios*. S.l., 245.

De Morgen (1997), *Vrijheid is onvermijdelijk, maar niet te snel*, 8/11, p. 70 (Freedom is unavoidable, but not too fast).

De Sola Pool, I. (1983), *Technologies of Freedom*. MIT Press, Boston.

Denham, A. (1996), *Think-Tanks of the New Right*, Aldershot, Dartmouth.

Dimech, P. & Turpin, E. (1998), 'Le réseau internet est-il économiquement viable', *Communications & Stratégies*, n° 32, pp. 23-66.

Doern, B. (1996), 'Towards an International Antitrust Authority? Key Factors in the internationalization of Competition Policy', *Governance*, Vol. 9, n° 3, pp. 265-286.

Dolfeyn, M. & Meert, B. (1996), 'Het gebruik van Telematica in Belgische KMO's', Aanvulling op Tweede Wetenschappelijk Verslag. (The use of telematics in Belgian SMEs)

Dordick, H. (1990), 'The origins of universal service: History as a determinant of telecommunications policy'. *Telecommunications Policy*, Vol. 14 n° 3, pp. 223-231.

Drucker, P.F. (1993), *Post-Capitalist Society*, Harper Business Books, New York.

Ducatel K., Fahrenkrog G. & Gavigan J. (1997), *Made for Living? Sustainable Welfare and Competitiveness*, IPTS report n° 15, IPTS, Sevilla, pp. 29-37, see also http://www.jrc.es/iptsreport/

Ducatel, K. (1998), 'Policy Visions 2005-2010', IPTS-paper presented at IUAP-seminar *The Information society and employment*, Liège, 14/05.

Dutch National Action Plan (1994), 'Information Superhighway - From Metaphor to Action'. Action programme, http://info.minez.nl/docs/nap-en.htm

Dutton W.H., Blumler J.E. & alii. (1994), 'The Information Superhighway: Britain's Response', *PICT Policy Research Paper*, n° 29, London.

Dybkjaer, L. and Christensen, S. (1994), 'Info-Society 2000, Ministry of Research and Information Technology' (Danish plan for the Information Society)

Dyson, K. and Wills, K. (eds) (1983), *Industrial Crisis. A Comparative Study of the State and Industry*, Robertson, Oxford.

EITO (1998), 'European Information Technology Observatory 1998'.

Elchardus, M. (1994), *Op de ruïnes van de waarheid*. Kritak, Leuven. (On the ruins of the truth)

Esping-Andersen, G. (1990), *The Three Worlds Of Welfare Capitalism*, Polity Press, Cambridge.

Esping-Andersen, G. (1996), (eds) *Welfare States in Transition: National Adaptations in Global Economies*, Sage, London.

Euromedia Research Group (1986), *New media Politics*. Sage, London.

Euromedia Research Group (1992), *Dynamics of Mediapolitics*. Sage, London.

Euromedia Research Group (1998), *Mediapolicy*. Sage, London.

European Commission (1996a), 'Learning in the Information Society: Action Plan for a European Education Initiative (1996-98)'. Brussels, Communication to the European Parliament, the Council, the Economic and Social Committee and the Committee of the Regions (Communication presented by Mrs Cresson and Mr Bangemann, in agreement with Mr Flynn).

European Commission (1996b), 'Communication to the European Parliament, the Council, the Economic and Social Committee and the Committee of the Regions on Universal Service for Telecommunications in the Perspective of a Fully Liberalised Environment'. COM(96) 73, 13 March, Brussels.

European Commission (1996c), 'Proposal for a European Parliament and Council Directive on the application of open network provision (ONP) to voice telephony and on universal service for telecommunications in a competitive environment (replacing European parliament and council Directive 95/62/EC)', COM(96) 419 final, 11 September 1996.

European Commission (1996d), 'Commission Communication on Assessment Criteria for National Schemes for the Costing and Financing of Universal Service in Telecommunications and Guidelines for Member States on Operation of such Schemes'. European Commission, COM(96) 608, Brussels.

European Commission (1996e), 'Networks for People and their Communities: Making the Most of the Information Society in the European Union - Supplement containing Working Groups' Reports'. First Annual Report to the European Commission from the Information Society Forum, Brussels.

European Commission (1996f), 'Green Paper - Living and Working in the Information Society: People first', Com(96) 389 final; EC, Brussels.

European Commission (1996g), 'Communication on the criteria of evaluation for national systems for the calculation of costs and the financing of the universal telecommunications service, and the directive guidelines for the member States in that which concerns the functioning of such systems. COM(96) 608, 27.11, Brussels.

European Commission (1996h), 'Communication of the Commission: Services of General Interest in Europe'. COM (96) 443 final, Brussels.

European Commission (1997a), 'Building the European Information Society For Us All', Final policy report of the high-level expert group, DGV for Employment and Social Affairs, Office for Official Publications of the European Communities, April. Luxembourg.

European Commission (1997b), 'New Technologies and Services: Current Position and Outlook', Brussels.

European Commission (1997c), 'Status Report on European Union Telecommunications Policy', 7 May, http://www.ispo.cec.be/infosoc/legreg/tcstatus.html.

European Commission (1997d), 'Green Paper on the convergence of the telecommunications, media and information technology sectors, and the implications for regulation - towards and information society approach' COM(97) 623, 3 December.

European Commission (1997e), 'Commission Communication concerning the Review under competition Rules of the Joint Provision of Telecommunications and Cable TV networks by a single operator and the abolition of restrictions on the provision of cable TV capacity over telecommunication networks', Directorate General IV, Brussels.

European Commission (1998a), 'Status of Voice Communications on Internet Under Community law and, in Particular, under Directive 90/388/EEC', OJ n° C 6, 10 January.

European Commission (1998b), 'Third report on the implementation of the EU telecommunications regulatory package', DN:IP/98/165, 18 February, Brussels.

European Commission (1998c), 'Directive 98/10/EC of 26 February 1998 concerning the application and provision of an open network (ONP) for voice telephony and the establishment of a universal telecommunications service in a competitive environment', *JOCE*, n° L 101/24 of 1 April1998, Brussels.

European Parliament (1997), 'Directive 97/33/EC of the European Parliament and the Council of 30 June 1997 relating to interconnection in the telecommunication sector, with a view to assuring a universal service and interoperability through the application of open network provision principles (ONP), JOCE, n° L 119/32 from 26.07.1997.

Eurostat, ITU, OECD (1995), 'Communication indicators for Major Economies 1995'.

Falch, M. (1997), 'Cost and Demand Characteristics of Telecom Networks' in: Melody, W. (ed.): *Telecom Reform: Principles, Policies and Regulatory Practices.* Technical University of Denmark, Lyngby: pp. 107-122.

Federal Communications Commission (1996), 'Recommended Decision in the Matter of Federal-State Joint Board on Universal Service, CC Docket n° 96-45'. Washington DC, Federal Communications Commission, 8 November.

Ferguson, M. (1992), 'The Mythology about Globalization', *European Journal of Communication*, Vol.7 n° 1, pp. 69-93.

Ferné, G. (1996), 'Information and Communication Technologies Infrastructures. Introduction' in: OECD (ed) *Employment and growth in the Knowledge-based Economy*. Paris, OECD, pp. 95-100.

Ferrera, M. (1996), 'The 'Southern Model' of Welfare in Social Europe', *Journal of European Social Policy*, Vol. 6 n° 1, pp. 17-37.

FET (1997), 'Proximus klopt Mobistar in tweede halfjaar 1997', *Financieel-Ekonomische Tijd* (http:/www.tijd.be/tijdnet/tbe/front/art/7513399.htm). (Proximus beats Mobistar in the 2nd quarter of 1997)

Flammé, M-A. (1989), *Droit Administratif*, I, Bruylant, Brussels.

Flichy, P. (1991), *Une histoire de la communication moderne. Espace public et vie privé.* La Découverte, Paris.

Foreman-Peck, J., & Millward, R. (1994), *Public and Private Ownership of British Industry 1820-1990.* Clarendon Press, Oxford.

Freeman, C. and Soete, L. (1997), *The Economics of Industrial Innovation*, Third Edition, Pinter Press, London.

Freeman, C. (1987), *Technology Policy and Economic Performance: Lessons from Japan*. Pinter, London & New York.

Fröbel F., Heinrichs J. & Kreye, O. (1986), *Umbruch in der Weltwirtschaft*, Rowohlt Taschenbuch Verlag, Reinbeck bei Hamburg.

Gabel, R. (1967), *The Development of Separations Principles in the Telephone Industry*. Michigan State University Press, East Lansing.

Gabel, R. (1969), 'The Early Competitive Era in Telephone Communication, 1893-1920', Law and Contemporary Problems, n° 34, pp. 340-359.

Galal A., Jones L., Tandon P., Vogelsang I. (1994), 'Welfare consequences of selling public enterprises - an empirical analysis', *Worldbank Studie*, Oxford University Press, New York.

Galbraith, J.K. (1992), *The Culture of Contentment*. Penguin, London.

Garnham, N., Joosten, M., & Owen, J. (1994), 'Le développement du service public dans les télécommunications et la radiodiffusion au Royaume-Uni'. *Réseaux*, 66 (Juillet-Aout), pp. 37-66.

Garnham, N. (1997), 'Universal Service'. In Melody, W. H. (Ed.): *Telecom reform: principles, policies and regulatory practices*. (pp. 207-212). Den Private Ingeniorfond, Technical University of Denmark, Lyngby.

Garnham, N. (1998), 'Information society theory as ideology: a critique'. *Loisirs et Société*, Vol. 21 n° 1, pp. 97-120.

George, V. & Page, R. (eds) (1995), *Modern Thinkers on Welfare*, PrenticeHall/Harvester Weatsheaf, Hertfordshire.

Geuna, A. and Steinmueller, W. E. (1997), 'Joining the Information Society: Internet Access Issues for Europeans', *FAIR Working Paper n° 17*, BETA, Université Louis Pasteur Strasbourg, MERIT, Maastricht University, and Science Policy Research Unit University of Sussex, February.

Giddens, A. (1990), *The Consequences of Modernity*. Polity Press, Cambridge.

Giddens, A. (1994), *Beyond Left and Right: The Future of Radical Politics*, Polity Press, Cambridge.

Gilster, P. (1997), *Digital Literacy*, Wiley, New York.

Gourevitch, P. (1986), *Politics in Hard Times. Comparative Responses to International Economic Crises*, Cornell University Press, Ithaca.

Graham S., Cornford J. & Marvin S. (1996), 'The socio-economic benefits of a universal telephone network. A demand-side view of universal service', *Telecommunications Policy*, Vol. 20 n° 1, pp. 3-10.

Graham, C., Lewin, D., Milne, C., Moroney, J. and Skouby, E. (1996), *The Consumer in the Information Society*. Ovum Ltd, London.

Group of Lisbon (1994), *Limits to competition*. VUBPRESS, Brussels.

Groupe de Lisbonne (1995), *Limites à la compétitivité - Pour un nouveau contrat mondial*, Editions La Découverte, Paris.

Grupp, H. & Linstone, H. (1999), 'National Technology Foresight Activities Around the Globe: Resurrection and New Paradigms', *Technology Forecasting and Social Change*, LX, 1 (Jan. 1999), pp. 85-94.

Haddon, L. (1998a), 'New Dimensions of Social Exclusion in a Telematic Society', *FAIR Working Paper* n°. 45, University of Sussex, March.

Haddon, L. (1998b), 'The Control of Communications: Imposing Limits on Telephony', *A report prepared for Telecom Italia*, University of Sussex, Brighton.

Handy, C. (1990), *The Age of Unreason*, Harvard Business Press, Boston.

Haque, M.S. (1996), 'Public Service Under Challenge in the Age of Privatisation', *Governance*, Vol. 9, n° 2, pp.186-216.

Hayek, F.A. (1960), *The Constitution of Liberty*, University of Chicago Press, Chicago.

Henisz, W. & Zelner, B. (1997), 'The Institutional Determinants of Infrastructure Growth in Telecommunications: A Cross-National Study'. *Paper presented at the 25th annual Telecommunications Policy Research Conference*. Alexandrai, Virginia, 27-29 Sept 1997.

Henrotte, V. e.a. (1996), 'Utilisation des Systèmes d'échange d'Information inter-Organisationnels (SIO) par les P.M.E. en Belgique francophone', *Actes du deuxième colloque international de management des reseaux d'entreprise* (CIMRE'96), Vol. 1.

Henten, A. and Thomas W. (1996), 'Danish telecommunications: keeping the policy options open', *Telecommunications Policy*, Vol. 20, n°. 9, pp. 669-684.

Hills, J. (1993), 'Universal Service: a Social and Technological Construct'. *Communications & Strategies*, n° 10, pp. 61-83.

Hirst, P. & Thomson, G. (1995), 'Globalisation and the Future of the Nation State', *Economy and Society*, Vol. 24 n° 3, pp. 408-442.

Hotelling, H. (1929), 'Stability in Competition', in: *Economic Journal*, Vol. 39, pp. 41-57.

Hudson, H. (1994), 'Universal Service in the Information Age'. *Telecommunications Policy*, Vol. 18 n° 8, pp. 658-667.

Hulsink, W. (1996), *Do Nations Matter in a Globalizing Industry? The Restructuring of Telecommunications Governance Regimes in France, the Netherlands and the United Kingdom (1980-1994)*, Eburon, Delft.

Hulten, O. (1984), *Mass Media and State Support in Sweden*, Sweden Books, Stockholm.

Hutchins, R., (1947), *A Free and responsible Press: Commission on Freedom of the Press*. University of Chicago Press, Chicago, Ill.

Ietto-Gillies, G. (1996), 'Globalisation and Integration, Analysis, Measurement and Policy Issues', *TSER Working Paper* n° 9.

Ingenieursblad (1997), 'Telecommunicatie in Vlaanderen: Wat nu?', *Het Ingenieursblad* (11-12). (Telecommunications in Flanders; what now?)

Innis, H. A. (1950), *Empire and Communication*. Oxford University Press, Toronto.

Interview Yves Blondeel, T-Regs, 14/11/97.

Interview Luc Mabille, BIPT, 21/11/97.

Interview Eric Van Heesvelde, BIPT, 5/12/97.

Jaeger, B. and Hansen, F. (1999), 'Multimedia in Denmark: An Overview' SLIM National Report, in Williams, R. and Slack, R. (eds): *Europe Appropriates Multimedia. A study of the national uptake of multimedia in 8 European Countries and Japan*, University of Trondheim, Senter for Teknologi of Samfunn (STS), rapport n° 42, pp. 83-122. Also available on the Web: http://www.ed.ac.uk./~rcss/SLIM/public/national/DK.html.

Jessop, B. (1990), 'Regulation Theories in Retrospect and Prospect', *Economy & Society*, Vol. 19 n° 2, pp. 153-216.

Jessop, B. (1994), 'The Schumpeterian Workfare State', in Burrows, R., Loader, B. (eds): *Towards a Post-Fordist Welfare State*, Routledge, London.

Jessop, B. (1995), 'The Regulation Approach, Governance and Post-Fordism: alternative perspectives on economic and political change', *Economy & Society*, Vol. 24 n° 3, pp. 307-333.

Jones, C. (ed.) (1993), *New Perspectives on the Welfare State in Europe*, Routledge, London.

Joosten, M. (1991), 'De Engelse Les, De Grenzen van Liberalisering van Telecommunicatie' (The English lesson, The limits of telecommunications liberalisation), *Informatie en Informatiebeleid*, Vol. 9 n° 2, pp. 31-36.

Jouët, J. & Coudray, S. (1990), 'New Communication Technologies: Research Trends'. UNESCO, Paris. (Reports and Papers on Mass Communication # 105).

Kahin, B. & Wilson, E. J. III (eds) (1997), *National Information Infrastructure Initiatives - Visions and Policy Design*, The MIT Press, Cambridge/Massachusetts and London/England.

Kalkman J.W., Hins A.W. & Jurgens (1995), ECM, *Communicatie- en Informatievrijheid in het Digitale Tijdperk*, Tjeenk Willink, Zwolle (Communication and Information Freedom in the Digital Age).

Kellner, D. (1997), 'New technologies, the Welfare State and the Prospects for Democratization', *Paper presented at the EURICOMM/SMIT Conference*; 'Communication, Citizenship & Social Policy: Rethinking the limits of the Welfare state'. Boulder (Colorado), 2-4/10.

Korten, C. (1995), 'A Financial System Out of Control', in *Third World Guide 1995/1996*, Instituto del Tercer Mundo, Montevideo.

Krugman, D.M. and Reid, L.N. (1980), 'The 'Public Interest' as Defined by FCC Policy Makers', in: *Journal of Broadcasting*, Vol. 24 n° 3, pp. 311-325.

Kubicek H., Beckert B., Williams R. Stewart, J. (1999b), *The Social Shaping of Multimedia in an International Perspective. A four Countries Comparison*, to be published in 1999.

Kubicek H., Dutton W.H. & Williams R. (eds) (1997), *The Social Shaping of Information Superhighways. European and American Roads to the Information Society*, Campus Verlag/St. Martin's Press.

Kubicek H., Schmid U. and Beckert B. (1999a), 'The development of Multimedia in Germany', SLIM National Report, to appear in Williams, R. and Slack, R. (eds): *Europe Appropriates Multimedia. A study of the national uptake of multimedia in 8 European Countries and Japan*, University of Trondheim, Senter for Teknologi of Samfunn (STS), rapport n° 42, pp. 157-214, Also available on the Web: http://www.ed.ac.uk./~rcss/SLIM/public/national/DE-Report.html

Lanham, R. (1995), 'Digital Literacy', *Scientific American*, Vol. 273 n° 3, pp. 198-200.

Lehmbruch, G. (1982), 'Introduction: Neo-Corporatism in Comparative Perspective', in Lehmbruch, G. and Schmitter, P.C. (eds): *Patterns of Corporatist Policy-making*, Sage, London, pp. 1-28.

Lehmbruch, G. (1984), 'Concertation and the Structure of Corporatist Networks', in Goldthorpe, J.H. (ed): *Order and Conflict in Contemporary Capitalism*, Clarendon Press, Oxford, pp. 60-80.

Lensen, A. (1996), 'The emergence of a European regulatory authority for telecommunications,' The European Parliament, Director General for Research, Luxembourg.

Levy, B.H. (1994), *La Pureté Dangeureuse*. Grasset, Paris.

Ling, R. (1997), 'One can talk about common manners!, The use of mobile telephones in inappropriate situations', in L. Haddon (ed.) *Communications on the Move: The Experience of Mobile Telephony in the 1990s*. European Commission, COST248 report of the mobile phone group.

Lipsey, R.G. (1995), *A Structuralist View of Technical Change and Economic Growth*, Reprint n° 38, Canadian Institute for Advanced Research, Toronto.

Littlechild, S. C. (1983), 'Regulation of British Telecommunications: Profitability, *Report to the Secretary of State*'. HMSO, London.

Lobet-Maris Cl., Pierson J., van Bastelaer B. in collaboration with Burgelman J.C., Neuckens F., Punie Y. (1999), 'Multimedia in Belgium', SLIM National Report, in Williams, R. and Slack, R. (eds): *Europe Appropriates Multimedia. A study of the national uptake of multimedia in 8 European Countries and Japan*, University of Trondheim, Senter for Teknologi of Samfunn (STS), rapport n° 42. pp. 21-82. Also available on the Web: http://www.info.fundp.ac.be/~cita/slim/Belgium.html.

Lobet-Maris, C. & van Bastelaer, B. (1996), 'Belgium, a Laboratory for Europe: The Federal Model and the Information Society', *Telematics & Informatics*, Vol. 13 n° 2/3, pp. 81-96.

Lodge, G.C. and Vogel, E.F. (eds) (1987), *Ideology and National Competitiveness: An Analysis of Nine Countries*, Harvard Business School Press, Boston.

Lowette, T. (1995), *The electronic information industry in Belgium- Supply, demand and market infrastructure*. GRID, Brussels.

Mansell, R. & Wehn, U. (eds) (1998), 'Knowledge Societies, Information technology for Sustainable Development', United Nations Commission on Science and Technology for Development, Oxford University Press, New York.

Mansell, R. and Silverstone, R. (eds.) (1996), *Communication by Design: The politics of information and communication technologies*, Oxford University Press, Oxford.

Mansell, R. (1993), *The New Telecommunications, A Political Economy of Network Evolution*, Sage, London.

Mansell, R. (1997), 'Consumer and Citizen Rights and Expectations: Commercialising Advanced Communication Services in the Information Society', *FAIR Working Paper n° 25*, Science Policy Research Unit, University of Sussex, March.

Marshall's, T. H. (1950), *Citizenship and Social Class*, Cambridge University Press, Cambridge.

Marvin, C. (1988), When Old Technologies were New: *Thinking about Communications in the Late Nineteenth Century*. Oxford University Press, Oxford.

McQuail, D. (1992), *Media Performance: Mass Communication in the Public Interest*, Sage Publications, London.

Melody, W. (1997c), "Interconnection: Cornerstone of Competition" in: Melody, William (ed.): *Telecom Reform: Principles, Policies and Regulatory Practices*. Technical University of Denmark, Lyngby: pp. 53-66.

Melody, W. (1997d), "Policy Objectives and Models of Regulation" in: Melody, William (ed.): *Telecom Reform: Principles, Policies and Regulatory Practices*. Technical University of Denmark, Lyngby: pp.13-28.

Melody, W. (ed) (1997a), *Telecom Reform, Principles, Policies and Regulatory Practices*, Den Private Ingeniorfond, Technical university of Denmark, Lyngby.

Melody, W. (1997b), 'On the Meaning and Importance of 'Independence' in Telecom Reform', *Telecommunications Policy*, Vol. 21 n° 3, pp.195-199.

Metcalfe, J.S. & Georghiou, L. (1998), 'Equilibrium and Evolutionary Foundations of Technology Policy', OECD STI Review 22, pp. 75-100.

Meyer-Krahmer, F. (1996), "Dynamics of R&D-Intensive Sectors and Science and Technology Policy", in OECD (ed): *Employment and Growth in the Knowledge-based Economy.* Paris: OECD, pp. 213-236.

Ministerie van Economische Zaken (1995), 'Economic policy, technology and growth'. Discussion paper 9502. The Hague.

Morin, E. (1987), *Penser l'Europe.* Gallimard, Paris.

Mueller, M. (1993), 'Universal Service in Telephone History', *Telecommunications Policy*, Vol. 17 n° 5, pp. 352-369.

Mueller, M. (1994), 'Le service universel dans l'histoire du téléphone: une reconstruction'. *Réseaux*, 66, pp. 13-35.

Mueller, M. (1997a), *Universal Service: Competition, Interconnection, and Monopoly in the Making of the American Telephone System.* The MIT Press and The AEI Press, Cambridge MA and Washington DC.

Mueller, M. (1997b), *Universal Service" and the New Telecommunications Act: Mythology Made Law*, Communications of the ACM, forthcoming http://www.vii.org/papers/cacm.htm

Narendran, P. (1997), 'Technology For Universal Service', *FAIR Working Paper n° 21*, Maastricht Economic Research Institute on Innovation and Technology (MERIT), University of Maastricht and Ovum Consultants, February, London.

NCMV (1991), 'De KMO's in het Belgisch bedrijfsleven'. *NCMV-cahier 1*, Brussel. (SMEs in the Belgian Economy)

Negroponte, N. (1995), *Being Digital*, Vintage Books, New York.

Neice, D. C. (1996), *Information Technology and Citizen Participation*, Department of Canadian Heritage, Ottawa, SRA 167, August.

Neice, D. C. (1997), 'Access to the Information Highway: Clarifications and Policy Options', mimeo, policy brief, Department of Canadian Heritage, Ottawa.

Neice, D. C. (1998), 'ICTs and Dematerialisation: Some Implications for Status Differentiation in Advanced Market Societies' *FAIR Working Paper n° 43*, Science Policy Research Unit, University of Sussex, March.

Neu, W. & Stumpf, U. (1997), 'Evaluating Compensation Requirements by Telecommunications Universal Service Providers: a New Challenge to Regulations'. *Communications & Strategies*, n° 26, pp. 165-182.

Nulens, G. & Van Audenhove, L (1998), 'An Information Society in Africa? An analysis of the Information Society policy of the World Bank, ITU and ECA'. Med Media Jemstone Network/Jordan Institute for Diplomacy/MacBride Roundtable: *Culture & Communication: A Global Information Society*, Amman, Jordan, 23-25 November.

Nussbaum, M. and Sen, A. (eds.) (1993), *The Quality of Life*, Clarendon Press, Oxford.

OECD (1990), 'Taxation and the International Capital Flows', OECD, Paris.

OECD (1991), 'Universal service and rate restructuring in telecommunications', OECD, Paris.

OECD (1993), 'Small and Medium-sized Enterprises: Technology and Competitiveness'. OECD, Paris.

OECD (1994), 'Science and Technology Policy: Review and Outlook'. OECD, Paris.

OECD (1995a), 'Impacts of National Technology Programmes.' OECD, Paris.

OECD (1995b), 'Communications Outlook 1995'. OECD, Paris.

OECD (1996a), 'Information Infrastructure Convergence', OECD, Paris.

OECD (1996b), 'OECD in Figures.' OECD, Paris.

OECD (1997a), 'Science, Technology and Industry: Scoreboard of Indicators 1997'. OECD, Paris.

OECD (1997b), 'Communications Outlook 1997'. OECD, Paris.

OECD (1997c), 'National Innovation Systems'. OECD, Paris.

OECD (1998), 'Work On Sustainable Development: A discussion paper on work to be undertaken over the period 1998-2001', OECD, Paris, see also http://www.oecd.org

OECD (1999), 'Communications Outlook 1999'. OECD, Paris.

OECD-TEP (1991), 'Technology in a Changing World'. OECD, Paris.

OECD-TEP (1992), 'Technology and the Economy: The Key Relationships'. OECD, Paris.

Office of Technology Assessment (1990), *Critical Connections: Communication for the Future*, Government Printing Office, Washington.

Office of Telecommunications (1995), *Universal Telecommunications Services: Consultative Document on Universal Service in the UK from 1997*. HMSO, London.

Office of Telecommunications (1998), *Oftel Sounds warnings about national grid for learning*, HMSO, London, 6 January, http://www.inside-cable.co.uk/n98q1aad.htm

Omhae, K. (1990), *The Borderless World*, Collins, London.

Owen, B.M. and Wildman S.S. (1992), 'Traditional Models of Program Choice', in: Owen, B.M. & Wildman, S.S., *Video Economics*, Harvard University Press, Cambridge, MA: pp. 64-100.

Pauwels, C. (1998), 'Integrating economies, integrating policies. The importance of antitrust and competition policies within the global audio-visual order', *Communications & Strategies*, Issue 30, pp. 101-132.

Petrella, R. (1997), *Het Algemeen Belang, Lof van de Solidariteit*, VUBPRESS, Brussel. (The general interest, ode to solidarity)

Picard, R. (1985), *The Press and the Decline of Democracy: the Democratic Socialist Response in Public Policy*. Greenwood Press, Westport, CN.

Pierson, Ch. (1991), *Beyond the Welfare State*, Polity Press, Cambridge.

Pierson, Ch. (1996), *The Modern State*, Routledge, London.

Pierson, P. (1994), *Dismantling the Welfare State?*, Cambridge University Press, Cambridge.

Polanyi, K. (1944), *The Great Transformation*, Beacon Press, Boston.

Pool, I. de Sola (1983), *Technologies of Freedom*, Harvard University Press, Cambridge MA.

Poot, A.P. e.a. (1997), *Innovatie in de regio: Provinciale innovatieprofielen*. Ministerie van Economische Zaken, Amsterdam. (Innovation in the regions, provincial innovationprofiles).

Porter, M. (1990), *The competitive advantage of nations*. Macmillan Press, London & Basingstoke.

Poullet, Y. (1996), 'Plaidoyer pour un ou des service(s) universel(s) d'information publique' in *L'accès à l'information publique : une clé pour la croissance économique et la démocratie électronique*', Stockholm Conference 1996, European Commission.

Quah, D. T. (1996), 'The Invisible Hand and the Weightless Economy', Centre for Economic Performance, *Occasional Paper* n° 12, Programme on National Economic Performance, London School of Economics and Political Science, April.

Raadgevend Comité voor de Telecommunicatie (1997), Derde jaarverslag, Brussel.

Rapp, L. (1996), 'Comment - Public Service or Universal Service?' *Telecommunications Policy*, Vol. 20 n° 6, pp. 391-397.

Rawls, J. (1971), *A Theory of Justice*, Clarendon Press. Oxford.

Rheingold, H. (1993), *The Virtual Community: Homesteading on the Electronic Frontier*. Addison-Wesley, Reading MA.

Rogers, E. (1983), *Diffusion of innovations*. (3°ed.) Free Press, New York.

Romer, P. (1993), 'Ideas and Things', *The Economist*, 11 September.

Rosenau, J. (1990), *Turbulance in World Politics, a Theory of Change and Continuity*, Princeton UP, Princeton.

Rosso, S.J. Del (1995), *The Insecure State: Reflexions on "the State" and "Security" in a Changing World*, Daedalus, Vol. 124 n° 2, pp. 175-208.

Royal Commission on the Press 1947-9, (1949), Report. Cmd. 7700. HMSO, London.

Ruigrok, W. & van Tulder, R. (1995), *The Logic of International Restructuring*, Routledge, London.

Samarajiva, R. (1997), 'Telecom Regulations in the Information Age', in Melody, W. (eds): *Telecom Reform, Principles, Policies and Regulatory Practices*, Den Private Ingeniorfond, Technical university of Denmark, Lyngby.

Schement, J. R. (1995), 'Beyond Universal Service: Characteristics of Americans without telephones, 1980-1993'. *Telecommunications Policy*, Vol. 19 n° 6, pp. 447-485.

Schiller, H. (1986), *Information and the Crisis Economy*, Oxford University Press, New York.

Schlesinger, P., Silverstone, R. & Corner, J. (eds) (1997), *International Media Research. A critical Survey*. Routledge, London.

Schmidt, V. (1995), *The New World Order Incorporated, The Rise of Business and the Decline of the Nation State*, Daedalus, Vol. 124 n° 2, pp. 75-106.

Schneider, V. (1997), 'Different roads to the Information Society. Comparing U.S. and European Approaches from a Public Policy Perspective', In Kubicek H., Dutton W.H. & Williams R. (eds): *The Social Shaping of Information Superhighways. European and American Roads to the Information Society*, Campus Verlag/St. Martin's Press, pp. 339-358.

Schoechle, T.D. (1995), 'Privacy on the Information Superhighway', *Telecommunications Policy*, Vol.19 n° 6, pp. 435-452.

Scott, J. (1996), *Stratification and Power: Structures of class, status and command*, Polity Press, Cambridge MA.

Sen, A. (1981), *Poverty and Famines*. Clarendon Press, Oxford.

Sen, A. (1982), 'Equality of What' (1979 Tanner Lecture at Stanford). In Sen, A.: *Choice, Welfare and Measurement*. Blackwell, Oxford.

Sen, A. (1995), *Inequality Re-examined*. Clarendon Press, Oxford.

Sen, A. (1997), *On Economic Inequality*. Clarendon Press, Oxford.

SERV (1996), *De sociaal-economische analyse van de telecommunicatie in de ruime zin van het woord*. SERV, Brussel.

Servaes, J. (1987), *Media Aid. Naar een 'ander' communicatie- en ontwikkelingsbeleid.* ACCO, Leuven (Media-aid. Towards another' communication and development policy).

Servaes, J. (1999), *Communication for Development. One World, Multiple Cultures*, Hampton Press. Creskill, NJ.

Sharp, M. (1989), 'Corporate strategies and collaboration: the case of ESPRIT and European electronics' in Dodgson, M. (ed): *Technology Strategy and the Firm: Management and public policy*', pp. 202-218.

Skouby, K.E. and Henten, A. (1998), 'The political economy of competition in telecommunication', *Communications & Strategies*, issue 30, 2nd quarter, pp. 199-214.

Steiner, P.O. (1952), 'Program Patterns and Preferences, and the Workability of Competition in Radio Broadcasting', in: *Quarterly Journal of Economics*, Vol. 66, pp. 104-122.

Steinmueller, W. E. (1992), 'The Economics of Production and Distribution of User-Specific Information via Digital Networks', in C. Antonelli (ed.) *The Economics of Information Networks.* North-Holland, Amsterdam, pp. 173-194.

Steinmueller, W. E. (1998), 'Virtual Communities Developments: Globally and in the ACTS and Telematics Application Programmes' *FAIR Working Paper n° 46*, MERIT, University of Maastricht (Netherlands), March.

Stewart, J. and Slack, R. (1999), 'Multimedia in the UK', SLIM National Report, in Williams, R. and Slack, R. (eds): *Europe Appropriates Multimedia. A study of the national uptake of multimedia in 8 European Countries and Japan*, University of Trondheim, Senter for Teknologi of Samfunn (STS), rapport n° 42, pp. 389-438. Also available on the Web: http://www.ed.ac.uk./~rcss/SLIM/public/national/UKfinal.html.

Stigler, G.J. (1971), 'The Theory of Economic Regulation', *Bell Journal of Economics*, n° 2.

Stiglitz, J.E. (1996), 'The Role of Government in Economic Development', in Bruno, M. and Pleskovic, B. (eds): *Annual World Bank Conference on Development Economics*, Washington DC.

Sussman, G. (1997), *Communication, Technology and Politics in the Information Age*, Sage, Thousand Oaks, CA.

Swank, D. (1998), 'Funding the Welfare State: Globalization and the Taxation of Business in Advanced Market Economies', *Political Studies*, Vol. 46 n° 4, pp. 671-692.

Théry, G. (1994), 'Les autoroutes de l'information'. *Rapport au Premier Ministre*, La Documentation Française, Paris.

Third World Guide 95/96, (1995) Instituto del Tercer Mundo, Montevideo.

Tunstall, J. & Palmer, M. (eds.) (1991), *Media Moguls*, Routledge, London.

UN (1996) *Demographic Yearbook 1995*. UN, New York & Geneva.

UN (1997), *Statistical Yearbook 1996*. UN, New York & Geneva.

UNCTAD (1995), *Technological Capacity-building and Technology Partnership: Field Findings, Country Experiences and Programmes*. UN, New York & Geneva.

UNCTAD (1996a), *Exchanging Experiences of Technology Partnership: The Helsinki Meeting of Experts*. UN, New York & Geneva.

UNCTAD (1996b), *New Technologies and Technological Capacity-building at the Enterprise Level: Some Policy Implications*. UN, New York & Geneva.

UNCTAD (1996c), *Emerging forms of Technological Cooperation: The case for Technology Partnership. Inner Logic, Examples and Enabling Environment.* UN, New York & Geneva.

Underhill, G. (eds) (1997), *The New World Order in International Finance*, MacMillan Press Ltd, London.

Underhill, G. (1991), 'Markets beyond Politics? The State and the Internationalisation of Financial Markets', *European Journal of Political Research*, n° 19, pp. 197-225.

Ungerer, H. (1988), *Télecommunications en Europe.* Perspectives Européennes, Bruxelles.

Ungerer, H. (1997), 'Open competition in the post-1998 European telecoms market', speech for the Swedish Telecoms Summit '97, 3 December, Stockholm.

Uzan, L. & Fontaine, G. (1998), 'Birth of a broadcast model on the Internet', *Communications & Strategies*, 30, pp. 155-170.

Van Den Brink, R.E.M., (1987) *Informatie over de informatie: handboek over de informatiemedia in Nederland 1938-1985.* Stenfert Kroese. Leiden. (Information about information: manual on information media in the Netherlands 1938-1985).

Van Lente, H. (1999), 'Multimedia in Netherlands: An Overview of Actors, Activities, And Issues', in Williams, R. and Slack, R. (eds): *Europe Appropriates Multimedia. A study of the national uptake of multimedia in 8 European Countries and Japan*', University of Trondheim, Senter for Teknologi of Samfunn (STS), rapport n° 42, pp. 305-332. Also available on the Web: http://www.unimaas.nl/~hvlente/overview.htm

Van Miert, K. (1997), *The Deadline of 1st January 1998: What does it mean for consumers of telecommunications services?*, http://europa.eu.int/en/comm/dg04/speech/seven/en/sp97063.htm

Vedel, Th.: (1997) 'Information Superhighway Policy in France: the End of High Tech Colbertism?', In Kahin, B. & Wilson, E. J. III (eds): *National Information Infrastructure Initiatives - Visions and Policy Design*, The MIT Press, Cambridge/Massachusetts and London/England, pp. 307-348.

Verhoest, P., & Vercruysse, J. P. (1994), 'The Role of the State in the Belgian Telecommunications Sector in the Nineteenth Century'. *Réseaux: The French Journal of Communication*, Vol. 2 n°1, pp. 93-111.

Verhoest, P., Vercruysse, J.-P., & Punie, Y. (1991), *Telecommunicatie en beleid in Belgie (1830-1991): Een reconstructie van de politieke besluitvorming van de optische telegraaf tot de oprichting van Belgacom.* Otto Cramwinckel Uitgever, Amsterdam.

Verhoest, P. (1994), 'Studiesyndicaat "Nieuwe Diensten op Kabel en Telefoonlijnen [Project Telenet Vlaanderen:] Deelopdracht regulering'. *Paper commissioned by the Gewestelijke Investeringsmaatschappij Vlaanderen en het ministerie van economie, werkgelegenheid en binnenlandse aangelegenheden van de Vlaamse Gemeenschap.*

Verhoest, P. (1996a), 'Naar een institutionele hervorming van de Belgische telecommunicatie-regulering'. *Study commissioned by Belgacom.*

Verhoest, P. (1996b), 'Openbare Telecommunicatie (1834-1994): De rol van de overheid in het netwerkbeheer'. *Proefschrift: Vrije Universiteit Brussel.* (Public Telecommunication 1834-1994: the role of government in network governance)

Vernon, R. (1979), *The Productcycle Hypothesis in a New International Environment*, Oxford Bulletin of Economics & Statistics, Vol. 41, pp. 225-267.

Waizer, M. (1983), *Spheres of Justice.* Basic Books, New York.

Wallerstein, E. (1984), *The Politics of the World-Economy: The States, The Movements and The Civilizations*, Cambridge UP, Cambridge.

Weber, M. (1978), *Economy and Society: Volume I*, Bedminister, New York.

Wilding, P. (1997), 'The Welfare State and the Conservatives'. *Political Studies*. Vol. 45 n° 4, pp. 716-726.

Williams, R. and Slack, R. (eds) (1999), *Europe Appropriates Multimedia. A study of the national uptake of multimedia in 8 European Countries and Japan*, University of Trondheim, Senter for Teknologi of Samfunn (STS), report n° 42.

Winsbury, R. (1998), *Why a pan-European regulator for telecoms is only a matter of time*, Intermedia, Vol. 26 n° 4, pp. 33-35.

Winseck, D. (1998a), *Reconvergence – a Political Economy of Telecommunications in Canada*, Hampton Press, Cresskill, N.J.

Winseck, D. (1998b), 'Pursuing the Holy Grail: Information Highways and Media Reconvergence in Britain and Canada', *European Journal of Communication*, Vol. 13 n° 3, pp. 337-374.

Witte, E. (1990), *Politiek en Democratie*, VUBPRESS, Brussels. (Politics and Democracy)

Worldbank Press Release: *Rethinking the State*, Washington, 25/06/1997, see http://www.worldbank.org/html/extdr/extme/1380.htm

Wyckoff, A. (1996), 'The Growing Strength of Services', *OECD Observer 200*, pp. 11-15.

Zacher, M. (1992) 'The Decaying Pilars of the Westphalian Temple', in Rosenau , J.: *Governance without Government, Order & Change in World Politics*, Cambridge UP, Cambridge.

Zsuzsa, F. (1996), 'The Free Market and the Welfare State', Dr. J.M. den Uyl-lecture, 9 December, see http://www.pvda.nl/diversen/denuyl/uyl96.html